About Island Press

Island Press is the only nonprofit organization in the United States whose principal purpose is the publication of books on environmental issues and natural resource management. We provide solutions-oriented information to professionals, public officials, business and community leaders, and concerned citizens who are shaping responses to environmental problems.

In 2003, Island Press celebrates its nineteenth anniversary as the leading provider of timely and practical books that take a multidisciplinary approach to critical environmental concerns. Our growing list of titles reflects our commitment to bringing the best of an expanding body of literature to the environmental community throughout North America and the world.

Support for Island Press is provided by The Nathan Cummings Foundation, Geraldine R. Dodge Foundation, Doris Duke Charitable Foundation, Educational Foundation of America, The Charles Engelhard Foundation, The Ford Foundation, The George Gund Foundation, The Vira I. Heinz Endowment, The William and Flora Hewlett Foundation, Henry Luce Foundation, The John D. and Catherine T. MacArthur Foundation, The Andrew W. Mellon Foundation, The Moriah Fund, The Curtis and Edith Munson Foundation, National Fish and Wildlife Foundation, The New-Land Foundation, Oak Foundation, The Overbrook Foundation, The David and Lucile Packard Foundation, The Pew Charitable Trusts, The Rockefeller Foundation, The Winslow Foundation, and other generous donors.

The opinions expressed in this book are those of the author(s) and do not necessarily reflect the views of these foundations.

About Forest Community Research

Forest Community Research is a nonprofit organization based in the northern Sierra Nevada mountains of California with satellite offices in Arcata, California, and Portland, Oregon. The organization is dedicated to advancing community well-being and community-based approaches to resource stewardship through research, education, and practice. Since 1993, the organization has worked to bridge the thinking of people and groups with different and at times competing ideas about social and natural resources.

COMMUNITY FORESTRY IN THE UNITED STATES

COMMUNITY FORESTRY IN THE UNITED STATES

Learning from the Past, Crafting the Future

Mark Baker and Jonathan Kusel

ISLAND PRESS
Washington • Covelo • London

Library of Congress Cataloging-in-Publication Data.
Baker, Mark, 1961–
Community forestry in the United States: learning from the past, crafting the future / Mark Baker and Jonathan Kusel.
p. cm.
Includes bibliographical references and index.
ISBN 1-55963-983-0 (cloth : alk. paper) — ISBN 1-55963-984-9 (pbk. : alk. paper)
1. Community forestry—United States. I. Kusel, Jonathan. II. Title.
SD565 .B35 2003
333.75'152'0973—dc21
2002015726

British Cataloguing-in-Publication Data available.

Printed on recycled, acid-free paper ♲

Design by Artech Group, Inc.

Manufactured in the United States of America
10 9 8 7 6 5 4 3 2 1

To the people who work to sustain communities and forests

Contents

Acknowledgments

This book is part of a project by Forest Community Research, with support from the Ford Foundation, to study community forestry in the United States. With a focus on community forestry's evolution and key challenges, the study sought to critically evaluate community forestry and contribute to its strategic effectiveness and success. Soon after the project's inception, an advisory group of community forestry specialists including professionals from government, nonprofit organizations, and academia was assembled. The advisory group provided guidance for this project at several critical junctures, beginning with a full-day meeting in Washington, D.C., in December 1999. At this meeting the group critiqued and offered suggestions regarding our study methods and general approach to the project. Over the next 2 years of research and writing, members of the advisory group were available as resources that we could rely on when necessary. Their commitment and contribution to this project culminated in a meeting in Taylorsville, California, in November 2001, where they reviewed a draft manuscript. The advisory group's sustained and extraordinarily high level of engagement with the material enabled us to significantly improve the final product. The advisory group included Beverly Brown of the Jefferson Center; Genevieve Cross, formerly with the Trust for Public Land and now an independent writer; Brian Donahue of Brandeis University; Gerry Gray from the Forest Policy Center of American Forests; Ed Marston of *High Country News;* Mary Mitsos, formerly with the Pinchot Institute and now with the National Forest Foundation; Shannon Ramsay of Trees Forever; and Steve Yaddof of the Forest Service. Michael Conroy and Jeff Campbell, both with the Ford Foundation, participated at different times in the life of the project. We deeply appreciate the generous contributions all the advisory committee members made to this effort.

We would also like to acknowledge the generous and insightful contributions of those we interviewed for the project and of those who participated in one of the three regional workshops held as part of the research effort. During the intervening 2-year period between the advisory group meetings, we conducted more than 55 interviews with key people within the community forestry movement. These people are listed in the Appendix. They are a diverse group. They include grassroots community forestry practitioners, civil servants, community forestry nonprofit staff members, private sector entrepreneurs, community organizers, woods workers, nontimber forest product gatherers, academics, and national environmental group and wood products industry representatives. In addition to these interviews, we held three regional workshops, one each in the Northeast, the Intermountain West, and the Pacific West. The purpose of each workshop was to bring together a diverse group of community forestry leaders to discuss an array of region-specific community forestry issues. The consistently high level of debate and discussion at these workshops reflects the high level of commitment of these people both to the goals of the community forestry movement and to the hard work and thinking necessary for their realization. The generous contributions to this project of those who participated in the workshops (also listed in the Appendix) were invaluable in revealing the complexities, challenges, and opportunities that the movement faces.

We would also like to thank Forest Community Research staff for the contributions they made to this project. Throughout this project Lorraine Hanson has provided excellent and invariably good-humored support. Joyce Cunningham, Will Kay, and Beth Rose Middleton each contributed to the project. Lee Williams was involved at the early stages of this effort and conducted several of the interviews. The efforts and contributions of all these people are gratefully acknowledged.

Appreciation is extended also to Donna House for her help in expanding the scope of this study. We are grateful also for the detailed and constructive comments of two anonymous reviewers. They helped provide conceptual and strategic clarity that improved this work.

Finally, we would like to acknowledge the financial support of the Ford Foundation and the William and Flora Hewlett Foundation. Their support of groups across the country has helped advance the principles and practice of community forestry. Their contributions to this effort made this book possible.

CHAPTER 1

The Landscape of Community Forestry

> A new approach to community development is in the making—one that asks people about the long-term needs of a place and of all its residents. We're in the process of building local institutions that take over the job of looking after public value on a volunteer basis, and we're learning how to reinvest in areas so that they'll be more valuable to the next generation than they are to ours. . . . I think we can now show that stewardship springs from connectedness—it gives people back a sense of thinking responsibly on behalf of the whole community, and it sends a shiver up the spines of the gatekeepers by reminding them that someone can take away their keys.
>
> —*Bob Yaro, former director of the Center for Rural Massachusetts (Hiss 1990:207–208)*

Across the United States people have taken up the challenging task of creating new relations between themselves and the forest ecosystems on which they depend. Their common goal is to improve the health of the land and well-being of their communities. Often, their efforts have arisen from desperate circumstances: political gridlock and intractable social conflict concerning forest management, local economic crisis resulting from reduced access to resources essential to a community's survival, and large-scale patterns of forest degradation and fragmentation that threaten the integrity of working forest landscapes. Seeking to reverse historical patterns of resource extraction that threaten ecosystems and weaken communities, practitioners and supporters of what has come to be called *community forestry* challenge the dominant paradigm of forest management. They reject continuation of the historical disenfranchisement of communities and workers from forest management. They critique the ways in which the practice of traditional science has not stewarded ecosystems and has privileged some at the expense of others. And they call for a stop to the all-too-pervasive trends of long-term disinvestment in ecosystems and human communities that have undermined the health of both.

To redress these shortcomings, practitioners of community forestry are developing a new approach and new ideas about restructuring relations between people and forests. A key tenet of this approach is the belief that sustaining forest ecosystems demands that forest communities and workers also be sustained. The twin objectives of healthy forests and healthy communities are not distinct; rather, they are two inseparable halves that together constitute a unity. One without the other is inherently unsustainable; only together can each be sustained. Realizing this vision of sustained forests and communities entails a radical reorientation of the ways in which democracy and science are practiced, markets and institutions influence patterns of disinvestment and investment, and resource management agencies mediate relations between government and society. These themes constitute some of the challenges and the promises of community forestry.

This book is a historically grounded analysis of the community forestry movement in the United States. It examines the current state of the field to assess where community forestry is now and where it might go in the future. This purpose is important for the same reason that community forestry is important: There is a broad consensus that the dominant paradigm of forest management bequeathed by the Progressive Era, with its associated bureaucratic and technocratic structures, has, for the most part, failed to steward forest ecosystems and maintain vital communities. Community forestry has emerged as an alternative or complementary model of forest management and therefore offers the promise of forest management regimes that may succeed where the progressive model has not.

Identifying the current state of community forestry and its potential future is also important because community forestry in the United States has reached a critical stage. No longer a series of spontaneous ignitions across the country, it has gained the coherence and profile of a national movement. In short, community forestry has become a force to be reckoned with. As one longtime supporter of community forestry recently remarked, "Community forestry is ready for take off." However, there remains considerable debate about the most desirable course for the movement to follow and even about which people and groups should be included, for not all those whose livelihoods depend on the forest ecosystem are part of the community forestry movement. Thus community forestry has reached a critical crossroads. This book is timely because part of its purpose is not only to reveal and clarify the nature of the crossroads but also to suggest and legitimate a trajectory, a method, and a process that in the long run are most likely to promote ecological stewardship and build healthy communities.

The National Backdrop of Community Forestry

Reinvigorating democratic institutions and fostering civic engagement are widely recognized as the biggest challenges of democracy in the United States today. This challenge has arisen as a result of the failure of the liberal democratic state to provide people with meaningful opportunities to participate in collective decision making regarding the economic, social, and environmental conditions that affect them. The prevailing structure of interest group–driven politics (known also as interest group pluralism) has produced a plethora of highly capitalized, centralized, and specialized political lobbying organizations that effectively advance their respective agendas at state and national levels. Through financial contributions individuals support the groups that promise to forward their interests. Battle lines harden as interest groups compete for funds and support. Government policy and actions result from the tense interplay of interest group politics and influence peddling on one hand and the ostensibly neutral scientist–expert advancing the interests of the public good on the other. Welfare programs based on trickle-down and income poverty alleviation are assumed to be adequate safety nets for those unable to prosper; other critically needed investments in community capacity building, their relationships with income poverty and environmental deterioration, and the concomitant variety of potential policy and investment responses are ignored.

Democratic participation and civic engagement are not the only casualties of the dominant American political economy. Impoverishment of communities and lingering or increasing environmental degradation symbolize the disruptive workings of capitalism and the limits of both trickle-down and centralized command-and-control environmental management and regulation. These trends stand in stark contrast with the strong economic growth of recent years, low unemployment rates, and spectacular wealth increases among some segments of society. They are also cause for concern given the general trends within state and federal government to privatize services and incorporate market-based models of government service delivery.

The overlapping spatial patterns of community decline and environmental degradation suggest that their causes, and therefore the possibility of their amelioration, may be linked. Furthermore, the historically weak political representation and civic engagement of such communities suggests that strengthened participatory planning processes and a more vibrant civic culture may be important components of a solution.

To many, these observations may sound trite. However, they are useful to note and reflect on because they have given birth to a family of community-based social movements, of which community forestry is one. These community-based social movements share much in common because the conditions they address arise from the same set of dominant political, economic, and social institutions, processes, and relationships. Given the common ground from which these movements have emerged, it comes as no surprise that they share many important attributes in terms of both the frameworks used to analyze constraints and opportunities and the strategies proposed and implemented to advance their causes. A brief review of some of the key features of two of these social movements establishes parallels with community forestry and points to the common warp and weft they share.

Civic Environmentalism, Sustainable Communities, and Community Forestry: Three Sister Movements

Two community-based movements, civic environmentalism and the sustainable communities movement, are closely related to community forestry.[1] A review of some of the key objectives and core concepts of these movements highlights the similar conditions from which community-based social movements emerge, their common challenges, and their shared approaches and principles.

Civic Environmentalism

Narrowly conceived, civic environmentalism concerns the potential for communities to partner with government in environmental protection and stewardship, particularly with regard to moving beyond traditional command-and-control environmental regulation and diversifying the array of policy instruments that are used to maintain or enhance environmental quality (John 1994). A broader, more encompassing interpretation of civic environmentalism focuses attention on the importance of "the civic capacity of communities to engage in effective environmental problem solving, and the relationship between the civic life of communities and environmental conditions" (Shutkin 2000:15). This interpretation informs the civic environmental movement and the wide variety of civic environmental projects, primarily located in urban areas, around the country.

The focus on the linkages between community building and environmental problem solving is a central tenet of this movement. When these two goals are considered in tandem, as integrated processes, they focus attention on democratic renewal and environmental protection or enhancement. Community building depends on strengthening civic democracy, founded on the premise that all citizens should be able to participate equally in the decisions and in the institutions that affect their lives. This notion of democratic participation emphasizes the importance of community-based decision making in which, through face-to-face deliberation, common purpose and common good can evolve. Civil society, social capital, and place, or the local environment, are the three constitutive elements of civic democracy, according to Shutkin (2000:31). Shutkin argues that the strength of civic democracy may be ascertained by examining the extent and nature of social capital, the degree of political participation, racial and socioeconomic equality, and the extent of public investment and privatization. These indicators of civic health also provide the basis for determining effective ways to strengthen communities and their environmental problem-solving abilities.

Shutkin (2000:128) suggests that civic environmental projects embody six core concepts: participatory processes, community and regional planning, environmental education, industrial ecology (reflecting the urban focus of civic environmentalism), environmental justice, and the importance of place. In any civic environmental project, to varying degrees, each of these core concepts is present; much the same could be said for most community forestry efforts. Here, we briefly dwell on participation and planning processes within civic environmentalism because of their close association with similar processes within community forestry. Participation within civic environmentalism involves face-to-face deliberations among all stakeholders to collectively craft mutually acceptable solutions to environmental problems and simultaneously strengthen and create community. In contrast to the traditional top-down expert-driven model of environmental problem solving, civic environmentalism empowers communities, with the help of experts, to devise their own solutions. Meaningful participation of this sort strengthens community-based decision-making capacities, enables citizens to monitor environmental problems, builds social capital and civic infrastructure, and facilitates productive collaboration with both the public and the private sector.

Civic environmentalism incorporates models of community and regional planning rooted in the work of regional planners concerned with the question of how to plan for sustainable communities. These models, originally developed by regional planners and thinkers such as Frederick Law Olmsted, Lewis Mumford, Benton MacKaye, and Jane Jacobs, embrace a systems approach to planning for community and environmental health. When combined with participatory processes, this approach to planning enables communities to identify the systemic issues that underlie and give rise to particular problems, devise long-term, comprehensive responses to those issues (which often include attracting and channeling investment for collective benefit), and engage in the important process of developing a shared vision of a community's future. Part of the planning process entails identifying information needs, strengthening feedback mechanisms, and monitoring changes over time. One innovative approach to developing community-based feedback mechanisms is the Community Indicators Network of the public policy group Redefining Progress. This civic science–oriented network uses community-based indicators of community health that stakeholders developed themselves to track trends, assess current conditions, prioritize actions and issues, and measure progress (Shutkin 2000:133). The process of developing and using community-based indicators strengthens community capacity and fosters the development of a collective vision of the future.

The Sustainable Communities Movement

The sustainable communities movement parallels civic environmentalism, and they both overlap community forestry along key dimensions. Civic environmentalism and the sustainable communities movement share the intellectual legacy of Patrick Geddes and Lewis Mumford, particularly with regard to the relationships between environmental quality, equity, and community well-being, the importance of place-based solutions to regional planning, the need for social cohesion and civic engagement, and the lack of faith in technological progress to solve pressing urban social and environmental issues. The sustainable communities movement is also centrally concerned with revitalizing democracy. Consistent with notions of bioregionalism and local self-reliance, also part of the movement, this concern often focuses on regional and local forms of democracy. Communitarianism and the community values it promotes, as illustrated by the grassroots communitarian movement of the late 1940s and recently revived by a number of scholars and policy makers (see Etzioni 1994, among others), is another important element of the contemporary sustainable communities movement.

Lamont Hempel (1999:51), in his review of the sustainable communities movement, suggests that it emerged out of "decades of frustration" by planners, local officials and business leaders, citizen activists, and environmental groups that resulted from their inability to manage growth in a socially and environmentally sustainable manner. The failure of traditional planning, zoning, and redevelopment instruments led to the search for different, community-based ways to steward the environment and support the growth of vibrant communities. Much like the systems approach of civic environmentalism, the sustainable communities movement incorporates interdisciplinary approaches that are based on the assumption that integrated solutions are needed to address contemporary environmental and social challenges. Hempel (1999:53) identifies four main orientations within this movement: a "capitals" framework approach, the urban design approach, the ecosystem management strategy, and the metropolitan governance orientation. Each has its own analytical focus, theoretical and applied questions, and set of sustainability indicators. Although all four orientations are interesting in their own right, only those that correspond closely to community forestry are discussed here.

The capitals approach to defining and achieving community sustainability is rooted in ecological economics. Initial formulations of natural capital were later expanded to include other types of capital such as human capital, human-created capital, social capital, and cultural capital (Viederman 1996). Within this formulation sustainability "is a community's control and prudent use of all forms of capital . . . to ensure . . . a high degree of economic security and achieve democracy while maintaining the integrity of the ecological systems upon which all life and all production depends" (Viederman 1996:46, quoted in Hempel 1999:55). This approach, though not without its critics, draws attention to the relationships between the various forms of capital, facilitates full-cost accounting, and emphasizes the importance of developing measures of the different types of capital—a prerequisite to any form of mutually beneficial exchange.

The ecosystem management orientation in the sustainable communities movement emphasizes ecosystem preservation and restoration as the overriding factor in community design and development in both urban and rural contexts. Hempel (1999:58) notes that sociopolitical factors have been underemphasized in this science-based management approach but that recently this imbalance has begun to be corrected. Furthermore, because of the scale associated with most ecosystem management efforts, their multijurisdictional nature, and their science-intensive monitoring and evaluation protocols, most ecosystem management initiatives have been initiated either by federal and state agencies or by large, national environmental groups. Therefore, they are not easily meshed with community-scale processes, although community participation is certainly an essential component of any successful attempt at ecosystem management.

The metropolitan governance orientation in the sustainable communities movement at first appears to be a somewhat contradictory mixing of scales. However, the underlying thrust behind this approach is the fact that communities are interdependent. They can exert both positive and negative influences on each other, and they are affected by nonlocal economic processes and global relationships. This underscores the need to connect local with nonlocal policy making and to develop regional governance frameworks to coordinate the interdependent effects of local communities' actions and to promote regional environmental quality and economic opportunity. Otherwise, as Hempel notes, "the goals of sustainable community end up looking parochial and selfish" (1999:61). Hempel suggests the concept of a "community of communities" as a possible vehicle for achieving intracommunity coordination and advancing community-based policies within state and national policy-making arenas.

Finally, as with civic environmentalism, community indicators are a central feature of the sustainable communities movement. These indicators of community sustainability assess economic, social, and ecological health; they are monitored to determine changes in direction and intensity. Given the emphasis on deliberative democracy within this movement, citizens must be involved in the development of indicators as well as monitoring. This can create problems if academics and professional analysts challenge the validity of community indicators. On the other hand, insisting on the use of community-based indicators can be an important step toward developing civic science.

Community Forestry

The degree of symmetry between the core components of community forestry and civic environmentalism and the sustainable communities movement is startling but not surprising. As we noted at the beginning of this chapter, these three movements are part of a family of community-based social movements that share the same warp and weft, although the specific features of their patterns are different. For example, although this book focuses primarily on community forestry in rural areas and extends to urban areas where rural–urban linkages and exchange relations are emerging, the goals of community forestry parallel those of the other two movements. In short, the objectives of the community forestry movement are to conserve or restore forest ecosystems while improving the well-being of the communities that depend on them. Although the connection between community well-being and forest ecosystem health may be more direct than in most civic environmental or sustainable community initiatives, the assumption that environmental and community health are interdependent links all three.

A useful way to frame the objectives of community forestry is through the triad of environment, economy, and equity. This triad can be conceived of as a three-legged stool; each leg is an essential component, necessary to ensure the stool's stability. Community forestry is an integrative enterprise that seeks to reorder relations between forest-dependent people and communities, between them and the wider political and economic systems with which they engage, and between them and the forests they depend on, in a manner that advances equity (especially within contexts of historically marginalized or disenfranchised communities) and promotes investment in both natural and community capital.

Core community forestry concepts parallel those of civic environmentalism and the sustainable communities movement. For example, community forestry practitioners emphasize the importance of participatory, collaborative, community-based decision-making and planning processes that include all the stakeholders likely to be affected by the forest ecosystem management plan or practice under consideration. Inherent in this notion of collaboration is the recognition that not all stakeholders have been involved in these planning processes and that eliciting their participation will take substantial investments of time, energy, and resources. The creation of new institutional relations between forest-dependent communities and the public agencies and industrial or nonindustrial owners that manage forests is another core community forestry concept. These new relations focus on the rights and obligations of communities with respect to forest resources and the importance of developing community-based participatory and civic science models of research, monitoring, and evaluation.

Investment is a central community forestry theme. One of the primary purposes of community forestry is to stem the flow of value from ecosystems and the communities whose well-being is tied to them. This purpose is achieved by integrating investments for forest ecological restoration with opportunities for local community revitalization. Integrative community-scale investments that promote equity, social justice, and forest health are the heart of the movement. When tied to the practices of forest management and ecosystem restoration, they give structure, form, and content to the otherwise abstract three-legged stool of environment, economy, and equity.

Methods and Organization

This book is the result of a study of community forestry. It is developed from a survey of secondary literature, interviews, workshops, and the authors' insights born of their association with community forestry. The roughly 60 semistructured interviews with community forestry practitioners and leaders from across the country were conducted in 2000 and 2001, along with three workshops held during the same period. To ensure regional representation, care was taken to interview people from regions where we were unable to hold workshops. The workshops, one each in Vermont, Colorado, and California, were designed to bring together a small number of leading community forestry practitioners and supporters for region-specific discussions of community forestry. The discussions focused on the current state of community forestry, barriers and opportunities, support needs, and strategies for overcoming current challenges. The structure of the interviews paralleled that of the workshops. The workshops and interviews inform the structure and content of chapters in the latter half of the book. The following paragraphs provide a preview of the chapters.

Chapter 2 discusses the historical antecedents of community forestry in the United States, specifically as they presage and inform the current community forestry movement. These early examples of community forestry include Native American forest management practices, traditions of Hispano community forestry in the Southwest, and early examples of community forestry in New England. The chapter also examines the work of key figures in the turn-of-the-century Progressive movement who argued forcefully that communities should participate directly in and benefit from the management of forest resources. Although the arguments of these key figures were eventually marginalized by the dominant technocratic and bureaucratic orientation of the Progressive Era, both early traditions of community forestry and the writings of the more socially minded members of the Progressive movement constitute a rich historical tradition of community-based forest management with contemporary relevance, one that in many respects presages important components of current community forestry initiatives.

Chapter 3 chronicles the evolution and dominance of the Progressive Era model of forest management, with a specific focus on the social and ecological ramifications of that model. In particular, we focus on the ways in which the dominant forest management regime separated community well-being from forest health and undermined work and occupation as a basis for forest enfranchisement. This chapter sets the broader context for understanding the rise of community-based forestry by examining the development of the conditions that led to its emergence.

Chapter 4 describes community forestry as a synergistic process involving simultaneous "ignitions" across the country at primarily local rural levels but almost always involving state and federal players as well. The emergence of community forestry is discussed as a response to the negative social and ecological outcomes of the dominant pattern of forest management. Community forestry is characterized as a process that seeks to reverse historical drawdowns of natural and community capital through reinvestment and redirection of benefit flows toward local groups who have previously not been a part of the broader political landscape of pluralistic political process. The conditions that gave rise to community forestry and the common themes that underlie its diverse regional forms are discussed, along with the organizations and some of the activities that led to its establishment as a social movement.

Chapter 5 outlines a framework for understanding and evaluating the goals of community forestry and analyzing the constraints and opportunities for advancing the movement. The triad of environment, equity, and economy is presented as a useful heuristic to capture the core of the integrative and overarching objectives of community forestry: the development of new relations between people and the forests on which their livelihood depends that maintain or enhance ecosystem processes, promote democratic values of civic participation and self-determination, and generate sustainable revenue streams for reinvesting in ecosystems and communities. The concepts of natural and community capital are presented as useful analytical devices for understanding community forestry. This framework also examines the operation and effects of the institutions that mediate between the various capitals and the relationship between natural and community capital and community well-being. The capitals framework and the environment, economy, and equity triad inform and guide the following chapters.

Chapter 6 evaluates the potential of community forestry to strengthen participatory structures and processes. The chapter begins with a critique of interest group liberalism. This leads to a discussion of the promising potential of community forestry to advance participatory democracy through the sorts of community-based deliberations and engagement associated with civic republicanism. Following this is a section that cautions against the wholesale embrace of civic republicanism because of equity and social justice concerns as they relate to forest workers and other less empowered groups. After a brief review of the philosophical and pragmatic reasons why equity and justice must be a primary objective of community forestry, the chapter addresses some of the ways in which the movement is currently grappling with equity issues. This leads to a discussion of strategies for promoting equity within community forestry. The final sections of the chapter discuss the linkages between local empowerment, participatory democracy, and community forestry.

Chapter 7 examines the implications of community forestry for the organization and operation of government. The role of the U.S. Forest Service in community forestry is evaluated, especially with regard to institutional capacity, the process of institutionalizing the "radical center" that community forestry represents, and the organization of work in the woods. Also addressed is the actual and potential role of extension forestry in community forestry. The chapter also examines the relationship between community forestry, national environmental groups, and the forest product industry. This includes a review and evaluation of the critiques some environmental organizations have made of community forestry and identification of possible areas of common interest.

Chapter 8 addresses the implications of democratization for the practice of science and the restoration ecology of community forestry. Community forestry offers a powerful critique of the dominant model of knowledge generation and acquisition, which is rooted in the Progressive Era model of technocratic bureaucracies and scientific expertise. In contrast to this model of traditional science, community forestry espouses a participatory model of knowledge production, one that integrates monitoring and adaptive learning, incorporates local knowledge, and empowers people. The differences between the traditional model of science and participatory civic science are explored, and case material is used to explore the on-the-ground effects of the differences between them.

Chapter 9 considers investment, one of the primary vehicles for achieving the goals of community forestry. This chapter begins with a brief review of the reasons why more value flows out of ecosystems and communities than back in and the investment strategies that have been developed to promote reinvestment in forest ecosystems. This leads to a discussion of the limitations of these approaches with respect to investing in community capital and of the need for investment strategies that explicitly empower people and improve community well-being. Although not always, such strategies often are linked to issues of social justice and equity. A case study is presented that illustrates the potential linkages between reinvestment in forest ecosystem health and increases in community well-being. The chapter also addresses more general issues concerning the challenges and opportunities associated with promoting community-scale forms of investment in natural and community capital.

Chapter 10 reflects on the potential of the community forestry movement to live up to its promises. It reviews the movement's transformative objectives and its implications for democracy, government, and science. It revisits the relationship between community forestry, environmental stewardship, and community well-being. It examines the potential of community forestry to strengthen civic society by promoting social engagement, through civic organizations, with environmental and socioeconomic issues. Finally, it revisits the linkages between community forestry and other contemporary community-based social movements.

CHAPTER 2

Historical Antecedents

> The footpath through New England's history of town-owned forests has turned full circle. In a sense, communities have begun to reclaim expansive common lands given up long ago to individual ownership. . . . Where land splinters once hewed from common domain and converted to severalty, public rights are now being reasserted through a similarly fragmented repertoire of local initiatives.
>
> —*McCullough (1995:301)*

The multicultural traditions of community-based forest management that once flourished in this country, along with the ideas of Progressive Era thinkers such as Benton MacKaye, present an astonishing array of community forestry models, practices, and ideals that together constitute important historical antecedents to the current community forestry movement. Although this historical legacy provides deep and stabilizing roots for the community forestry movement, the marginalization throughout most of the last century of these traditions and ideas warns us of the power and influence of forces that would prefer to see community forestry fail. Some of the main elements of this legacy consist of indigenous models of forest management, traditions of common water and forest management imported from the Old World and adapted to new circumstances in the Southwest and Northeast by early settlers, and the brief but significant moment during the Progressive Era when foresters such as Bernard Fernow, Samuel T. Dana, and Benton

MacKaye attempted to integrate community sustainability with the practice of forestry. The practices, goals, and objectives of these diverse historical antecedents of community forestry include community-based and participatory collective decision-making processes, communal resource ownership, the development of local knowledge and its use in sophisticated resource management regimes, explicit linkages between collective identity and resource stewardship, and the search for equitable, just, and stable institutional arrangements for managing public forests.

Throughout most of North America, Native American groups developed sophisticated forest management practices. These included a wide variety of management practices to promote desirable species and seral stages and the extensive use of fire as a management tool. Growing concern about the ecological effects of decades of fire suppression has renewed interest in reconstructing, learning from, and adopting aspects of these early examples of fire and forest management. In the Southwest, Hispano communities carried with them from Spain traditions of communal water and forest management. In the upper Rio Grande these traditions mingled with indigenous peoples and traditions. The correspondence between collective management, community identity, access to resources, and stewardship in the Southwest resonates with many of the core themes of contemporary community forestry. The deeply rooted tradition of municipal and town forests of New England, in many instances derived from much older European traditions carried to the New World by early settlers, constitutes a third historical legacy with direct relevance for contemporary community forestry. This legacy firmly establishes the urban roots of community forestry and explicitly links community forestry with civil society and the support of local public institutions. Finally, the Progressive Era, better known in forestry circles for its legacy of scientific forestry and hierarchical authority structures, also included a small group of foresters who promoted community-based forestry on public lands and who fervently worked to make the practice of forestry consistent with sustainable communities, strong local public institutions, and a vital civic culture. This diverse legacy informs, grounds, and inspires the current community forestry movement.

Although the dominant patterns of forest management that emerged in the early twentieth century marginalized and suppressed these early community forestry antecedents, key elements remain intact and inform current efforts to advance community forestry today. For example, in New England there is renewed interest in municipal and town forests, particularly as islands of green space standing against the rising tide of development and pervasive forest fragmentation. Municipal and town forests, along with the communitarian ideals of forest management that embody and the local institutional practices of town meetings in which they are embedded, constitute legacy social forests that can function as anchors for the spread of community-based forestry more widely across the landscape. As national debates over the most effective ways to reverse a century of fire suppression and resulting changes in forest ecosystem structure and function continue to gather momentum, the potential roles of rural communities and the existing workforce as partners in forest and fuels management are accorded increasingly high priority. Native American models of community-based forest management inform these debates and have direct relevance for forest management across the country. Hispano communities in northern New Mexico, one of the flash points for the emergence of contemporary community forestry, can teach other community forestry practitioners, supporters, and students about the strong linkages between community identity and the resources on which a group of people collectively depend. Finally, the visions of socially sustainable forestry espoused and advocated by carly foresters and progressives such as Bernard Fernow, Samuel T. Dana, and Benton MacKaye have direct relevance for current attempts to strengthen civic institutions in rural communities, improve community well-being, and address unresolved labor issues within the forestry sector.

An analysis of these antecedents illuminates and provides historical depth to many of the current concerns and issues of the current movement. Furthermore, considerations of how and why these antecedents were thrust aside by the dominant paradigm of forest management in the twentieth century, a theme taken up in Chapter 3, is instructive because it helps identify the key challenges that the current movement faces. Analyzing the causes of the marginalization of the historical antecedents of community forestry helps to illuminate the nature of the political, bureaucratic, science-based, and economic barriers that confront and hinder the realization of the goals and objectives of the current community forestry movement.

Native American Traditions of Environmental Management

The last two or three decades have witnessed dramatic shifts in our understanding of American Indian resource management practices. Contrary to the long-dominant view that the landscapes the early Anglo-European settlers encountered were wild and exclusively the results of "natural" processes (i.e., processes to which aboriginal populations were ancillary at best), it is now generally accepted that those "wild" pre–European settlement environments reflected in myriad ways enduring, complex, and sophisticated human activities designed to favor desired combinations and frequencies of plant and animal species. For example, it is now recognized that the open parklike forests of southern New England were the products of American Indian burning practices. Indians burned to clear land for planting crops; the remaining forests were frequently fired, often twice a year, to facilitate travel and to attract game. The result, at the landscape scale, was to create a mosaic of forest types at different successional stages, thus promoting the ecological edge effects that were attractive to game. This, in turn, raised the total population of herbivores as well as the predators that preyed on them. Indeed, as one scholar has commented, "Indian burning promoted the increase of exactly those species whose abundance so impressed English colonists: elk, deer, beaver, hare, porcupine, turkey, quail, ruffed grouse" (Cronon 1983:51).

Even in areas where agriculture was not practiced, hunter and gathering groups did not harvest the plants and animals that were simply "naturally" present. Rather, they too were involved in millennia-long processes of gradually domesticating the environment through a wide range of management practices. These practices, which included burning, pruning, propagating, digging, weeding, cultivating, and thinning, not only resulted in greater populations of desired species but also had widespread ecological effects. For example, with regard to Native Californian environmental management practices, Blackburn and Anderson (1993:19) state,

> Important features of major ecosystems had developed as a result of human intervention, and many habitats (e.g., coastal prairies, black oak savannas, and dry montane meadows) were deliberately maintained by, and essentially dependent upon, ongoing human activities of various kinds. In fact . . . the vertical structure, spatial extent, and species composition of the various plant communities that early European visitors to California found so remarkably fecund were largely maintained and regenerated over time as a result of constant purposive human intervention.

Thus, the natural environments that early European visitors and settlers encountered in many regions of North America were profoundly influenced by, and the product of, American Indian environmental management practices. However, most early Europeans seem to have been unaware of the role of American Indians in creating the natural environments they encountered. In this regard Nabhan (1997:160) asks, "Is it not odd that after ten to twelve thousand years of indigenous cultures making their homes in North America, Europeans moved in and hardly noticed that the place looked 'lived-in'?" Part of the answer may lie in the fact that American Indian environmental management practices generally involved the use of low-technology material implements. The simple material component of the indigenous environmental management technology has unfortunately led some commentators to dismiss the importance of American Indian environmental management. Others argue that this reflects an ethnocentric bias that prioritizes a materialistic view of technology and undervalues technology as a system of knowledge (Lewis 1993:395–396). In fact, Native American management practices were embedded in sophisticated knowledge systems that reflected a deep understanding of the ecological relationships involved. These knowledge systems were interwoven with cultural understandings of place, home, the use of plants and animals, and collective identity.

Important ecosystem components were maintained and regenerated by American Indian environmental management practices. With a lens on the Southwest, Nabhan states (1997:161–162),

> It can no longer be denied that some cultures have developed specific conservation practices to sustain plant populations of economic or symbolic importance to their communities. The O'odham are among those who still protect rare plants from overharvesting near sacred sites, who transplant individual cacti and tubers to more protected sites, and who once conserved caches of seeds in caves to ensure future supplies. They have pruned and promoted the fruiting of certain rare food plants, plants they claim to be rarer now than when Native Americans intensively used them.

The cessation of many American Indian environmental management practices has led to the gradual decline of the ecosystem components they previously maintained and regenerated. The positive association between use and abundance was and is widespread among American Indian gatherers. Thus Blackburn and Anderson (1993:19) note, "When elders today are asked why the rich resource base and fertile landscape that they remember as having existed in the past has changed so drastically, they are apt to respond by saying, 'No one is gathering anymore.'"

The correspondence between use and abundance has important contemporary resource management implications. Many currently rare, threatened, or endangered species are those that were previously abundant and used extensively by American Indian groups. The absence of historical patterns of human plant and animal use is, in some cases, a primary cause of their decline. Therefore, the return of ecosystems to their historic "presettlement" range of variation (a common goal of forest ecosystem restoration) and the conservation of many currently endangered plants and the ecosystem components on which they depend probably will entail the reintroduction of indigenous management practices. And because of the link between plant abundance, ecological knowledge, access, and cultural continuity, reintroducing indigenous management practices involves much more than attempting to replicate presettlement anthropogenic fire frequency and intensity, for example. Instead, land managers and scientists must acknowledge the importance and integrity of traditional ecological knowledge systems and collaborate with those who retain that knowledge.

Innovative efforts by some public lands agencies in California and elsewhere have begun to integrate traditional ecological knowledge into public lands management. Land managers in the Six Rivers National Forest in northern California have worked with Yurok, Karuk, and Hupa elders in designing experimental burns to promote the growth and quality of beargrass (*Xerophyllum tenax*), a culturally important plant species that needs firing to produce desirable young shoots. Redwood National Park resource managers, in consultation with Hupa basketweavers, have begun experimenting with management practices, including firing, that will promote and restore hazel (*Corylus cornuta*) stands, many of which still show evidence of past Indian management. And managers in the Sierra National Forest, in consultation with Indian elders, are monitoring the effects of simulated indigenous burning of deer grass (*Muhlenbergia rigens*) on growth and reproduction. After the cessation of Indian management of deer grass in the area, deer grass colonies that were once high-quality gathering sites have dwindled and declined. Burning deer grass colonies stimulates the growth of longer, thicker culms, increases flower stalk production, and may facilitate the establishment of new colonies. These efforts to integrate traditional ecological knowledge into public land resource management, though still in their early stages, are important because they represent efforts to simultaneously conserve and enhance otherwise dwindling plant species and their associated ecological niche and to maintain the knowledge base that "keeps the ancient harvesting and management traditions of indigenous cultures alive" (Anderson 1993:173–174).

Community Forestry and Collective Action in Southwestern Hispano Communities

Whereas traditions of municipal and town forests were transplanted from northern Europe to New England, in the Southwest Hispano communities molded the landscapes they found using forest and water management traditions they carried with them from Spain and Mexico. These communities have managed to retain their distinctive rural character, cultural integrity, and community identity despite long-term and widespread rural poverty, a harsh environment, and diminishing access to natural resources. In the seventeenth and eighteenth centuries Spanish colonists settled in the upper Rio Grande in present-day New Mexico. These early settlers brought with them traditions of communal resource management and systems of property rights from Spain. These people and their Old World traditions interacted with the preexisting indigenous Puebloan and nomadic Native American groups. As a result of this interaction, Hispano communities often included mestizo or mixed-blood people who became acculturated in Spanish customs and Christianity (Carlson 1990:159). In a similar manner some resource regimes, especially those concerning water management, expanded on preexisting indigenous systems using Old World social and technical design principles (Hutchins 1928, cited in Brown and Ingram 1987:49).

Access to land for Spanish settlers, especially during the resettlement of the region that began approximately 10 years after the successful Pueblo Rebellion of 1680, was guided by the system of land grants made to communities and individuals based on Spanish law and resting on the political authority of the Spanish government. Groups of Spanish settlers gained rights to land through the institution of the community land grant. The primary purpose of the community grants, which were made to groups of at least 10 families or 12 adults, was to encourage Spanish occupation of the region by the growing population and to secure the northern border areas against American Indian attacks and competing sovereign claims (Carlson 1990:23). Community land grants included irrigable bottomlands as well as large upland areas and combined private and communal forms of land tenure. Individual grantees were allotted strips of irrigable land for their private use, and the adjacent meadows, upland areas, and higher pinyon pine, juniper, and ponderosa pine forests within the grant were managed communally for pasturage, timber, game, and other forest products.

The irrigable bottomlands almost always comprised only a small fraction of the total area of the grant. They extended from the river to the abrupt break between the floodplain and upland areas. To ensure an equitable distribution of irrigable land, the system of long lots was instituted. Long strips of agricultural land, with a length to width ratio of at least 3:1, were oriented perpendicularly to the river and extended to the edge of the bottomland. The main communal irrigation canal (*acequia*) followed the contour of the break between bottomland and upland, thus providing water to the upper end of each long lot; water diverted from the canal would irrigate the lot, with return flows rejoining the river to be diverted again by a downstream *acequia*. Each allottee received one of these long lots, which ranged in size from 10 to 20 acres, thus ensuring access to irrigated agricultural land for all members of the community (Carlson 1990).

The communal irrigation systems (*acequias*) used to irrigate the bottomland areas of the community land grants, and the cooperative associations (also known as *acequias*) formed to manage them, are fascinating examples of community-based collective action; many still function today. All tasks associated with *acequia* management, including the original construction of the primary and secondary irrigation channels, the periodic reconstruction and repair of the diversion structure, and yearly channel maintenance, were and are performed through voluntary labor contributions from those whose lands the *acequia* irrigates. The mobilization of labor and task supervision, allocation of water, and conflict resolution during times of water scarcity are the responsibilities of the *mayordomo* (ditch boss), an institutional role that originated in Spain.

Acequias are core community institutions central to the Hispano communities of the upper Rio Grande. In addition to being recognized political entities, *acequias* and the water they provide in this arid and semiarid environment continue to play important roles in terms of Hispano community identity and cultural continuity. The multiple meanings of the term *acequias* illustrate the intertwined nature of the physical irrigation system and the social system organized to manage it. As Brown and Ingram note (1987:33, 55) the community value of the water *acequias* convey is as important as the commodity value. By this they mean that water *and* the social relations that govern and enable its use are central to the continuity and integrity of Hispano culture.

Brown and Ingram identify four elements that together comprise much of what people mean by "preserving Hispano cultural continuity." In addition to attachment to water, they identify preservation of the Spanish language, the importance of family relationships, and attachment to land. The attachment to the land that Hispano communities have owned privately and communally for more than three centuries runs deep and strong. It extends up from the irrigated bottomlands to include the adjacent meadows, upland areas, and forested mountains that were included in the original community land grants. These areas were important components of Hispano livelihood strategies because they provided pasturage for stock raising (primarily sheep but also to some extent cattle and draft animals), which was an important complement to irrigated agriculture. The size of herds and the number of stockmen were always significant, but they both increased dramatically in the nineteenth century. By the beginning of the twentieth century, stockmen outnumbered farmers in the peripheral regions of the Hispano homeland (Nostrand 1992:75).

Most of the lands Hispano communities used for grazing their extensive herds, meeting their needs for timber and other forest products, and hunting and fishing were located within the original community land grants that the Spanish government had given them. However, through a variety of unfortunate circumstances, by the mid-twentieth century most of these common lands had been lost or sold (Carlson 1990:94). The gradual decline of these common lands began immediately after the signing of the 1848 Treaty of Guadalupe Hidalgo. During the adjudication of Spanish and Mexican land grant claims by the U.S. government after the treaty's ratification, a portion of the communal land claims were not recognized as legitimate. Despite legal challenges to this decision, these lands were transferred to public ownership. Further reductions in the extent of Hispano common lands can be attributed to federal tax assessment policies. Because taxes were assessed communally on communal portions of the land grants, if one family was unable to pay their share of the tax burden or refused to pay taxes on principle, the communal area as a whole went into tax default status. Because of the economic hardships of the Depression, a large amount of communally owned land was lost in this manner. Most of it was transferred by the state of New Mexico to the Soil Conservation Service or the Forest Service. Additionally, during the Depression the federal government was authorized to purchase land from landowners who were in financial trouble. As part of this purchase program, the federal government purchased all or portions of the communal areas of many of the remaining patented Spanish and Mexican land grants. Most of this land was transferred to the U.S. Forest Service and today constitutes much of the Carson and Santa Fe National Forests. Hispano communities still resent this action and want to reassert their customary rights over these former communal areas.

After the assumption of Forest Service control over much of this communal land, a series of extremely contentious reductions occurred in the size and number of herds of stock that Hispanos could graze on the now "public" lands. Other customary uses of these areas such as woodcutting were restricted.[1] Hispano communities also resented Forest Service timber harvesting practices on former grant lands, especially when local communities saw no benefit and only harm from those activities. The Vallecitos Sustained Yield Unit was an attempt, albeit only partially successful, by the Forest Service to mitigate the negative effects of grazing reductions by attempting to provide secure forest-related employment to local Hispano communities (Krahl and Henderson 1998). Hispano resistance to the curtailment of their customary resource use and access rights took the form of violent and nonviolent actions such as roadblocks to prevent log trucks from leaving timber harvest areas, harassment of Forest Service personnel, and forest arson. By the 1980s, because of pressures in part from environmental groups, some portions of the Carson and Santa Fe National Forests were completely closed to woodcutting and other customary uses. The restrictions on customary use are primary contributors to the reemergence of the community forestry movement in the Southwest, a story that we pick up in later chapters.

Town and Municipal Forests: Community Forestry New England Style

New England has a long history of community-based forest management. From the early period of Anglo-American settlement to the twenty-first century, New England towns and villages have managed common forests for diverse purposes. These include meeting local needs for fuelwood, commercial timber production, water supply protection, and land reclamation. Community forests also supported local community institutions. Community forests and woodlots subsidized the salaries of town ministers, supported local schools, and were the basis for some town welfare efforts. Examples of community forests include municipal or town forests, watershed plantations, forest parks, and common forests (McCullough 1995).

Origins and Early Traditions of Communal Woodlands in New England

Early traditions of community forestry drew on diverse models of common land forest management that settlers brought with them from Europe. Groups who established communities in New England did so through grants, or charters, given to them by colonial governments that specified the terms and conditions under which the proprietors, as the founding group came to be known, could retain title to the areas within the village boundary. The earliest settlements often were located in areas that previously had been periodically burned by Native Americans to clear undergrowth. In this way, Native American forms of forest management strongly influenced the subsequent spatial pattern of Anglo-European colonization and its ecological impacts. The proprietors allocated individual house lots as well as parcels for community needs such as the meetinghouse, school, parsonage, cemetery, and central common area for livestock grazing and militia training. The remaining unallocated area within the village boundary consisting primarily of uplands, meadows, and wetlands was held in common, with each owner possessing a proportional right to its use (McCullough 1995:15). Portions of these common lands were defined for specific uses such as grazing or woodlots; the remainder was quickly privatized as population and land pressures increased. The areas designated as woodlots or communal woodland constituted the first markers of New England's tradition of communal forests.

These "community forests" were intimately linked with settlements; they were utilitarian in nature, satisfying a wide variety of community needs including "grazing, cultivation, and felling of timber and wood for . . . framing, lumber, clapboards, shingles, staves, fences, bark for tanning, pitch or fuel" (McCullough 1995:16). They were located in areas that eventually developed into urbanized spaces, and their use was controlled by the voting population of each community. Another form of community-based land ownership, known as stinting, occurred in areas called cow commons, cow pasture, common meadow, ox pasture, or common field. Although actual title to these meadows or pastures rested with individual proprietors, often in proportion to their private landholdings, in practice common cultivation, grazing, or removal of timber or stone was allowed during specified times of the year. In such situations pasturage rights were shared among town members, including commoners and noncommoners. Stinting rights were considered quite valuable and were offered as inducements to new settlers (McCullough 1995:23).

Throughout the seventeenth century demands on wood products increased dramatically as both domestic and export markets rapidly expanded. At the same time, town administration became more sophisticated. In response to localized wood scarcity, towns developed a wide array of regulations and taxes designed to control and restrict the extraction of wood products and to generate revenue from commercial uses of communal woodlands. Many restrictions emphasized the importance of minimizing waste and using forest products efficiently. Steadily growing populations led to conflicts between proprietors and newcomers over access and use rights in the communal woodlands. Privatization of these areas often was chosen as one way for groups to perfect their rights in communal forestlands, whether they be proprietors on their way to becoming a landed class or newcomers and "cottagers" securing their tenuous claims to land and forest resources. In either case the result was a decline in the extent of communal woodlands and a reorientation toward private rather than communal gain. Continuous trespassing on these formerly communal lands indicates that common rights were not always easily extinguished. This process accompanied the establishment of new settlements from roughly 1700 onward until most of New England was allotted into towns. The pattern of these subsequent settlements was less nucleated as the need for defense declined and individual farming operations were consolidated into single larger ownerships.

During this period an important distinction began to emerge between communal woodlands managed to help satisfy the utilitarian needs of settlements (primarily those of the proprietors) and communal woodlands, more appropriately defined as public lands, that were set aside for the long-term support of local community institutions. Examples of these types of community forests include areas set aside for the use of clergy and for millers who operated saw or gristmills. Many communities also set aside communal forests to subsidize the salaries of ministers, teachers, and other school costs such as books. The annual salaries of ministers often included a commitment on the part of the parishioners to cut and deliver winter cordwood each autumn from the minister's woodlot. These designated public forests were owned by the institutions they were intended to benefit (i.e., the church or school) or by the towns. Occasionally communal forests and the buildings they contained were used to house and care for the community's poor population. Often these "poor farms" were sold by towns in the late nineteenth century as state and county government assumed greater responsibility for public welfare. However, many were not. They continued to be managed for timber production, and they constitute an important contribution to the present acreage of town and municipal forests.

Management authority for these forests generally rested with individual stewards appointed by the town authorities (often called "tree wardens") or with the beneficiary institution itself. Forest management and stewardship activities on these forests constitute the earliest examples of nonindigenous community forestry practiced in the United States. Management activities included organizing and supervising timber auctions (this included specifying the terms and conditions of the timber sale such as requirements to burn slash from harvesting and sow grass seed after the harvest), supervising the rental agreements governing the lessee's use of the land, and, by the end of the nineteenth century, initiating reforestation activities. In some cases proceeds from these forests were invested by the local community and became important sources of revenue generation for municipal purposes in their own right. These early community forests also make up a large proportion of the town and municipal forests that exist today and constitute important buffers against widespread forest fragmentation and residential home construction.[2]

Early Forest Service Support for Community Forestry on Locally Owned Public Lands

Community forestry as an organized movement with government support gained momentum at the close of the nineteenth century, when public concern about forest depletion, destructive fires on cutover forestlands, and the fate of abandoned unproductive agricultural lands combined with the emerging science of forest management to create conditions ripe for the growth of the town forest movement in the late nineteenth and early twentieth centuries. The early professional foresters who promoted community forestry, though certainly motivated by traditional forestry concerns regarding commercial timber production, also were keenly aware of the links between community forestry and broader social issues related to town and regional planning and community development. The country's first professionally educated and trained forester and the head of the forestry division in the Department of Agriculture, Bernard Fernow, in 1890 proposed that a movement be inaugurated to create community forests in America based on the Germanic model of communal forest management. Drawing on the well-known example of the Sihlwald, Zürich's ancient city forest, Fernow described how town forests, if well managed, could yield both steady income and employment opportunities.

Federal government involvement in the support of community forestry was also advocated by two other foresters, Raphael Zon, a Russian-born forester and colleague of Gifford Pinchot in the Forest Service, and forester Samuel T. Dana. Both foresters presented papers on community forestry to the National Conservation Commission, convened in 1908 by Theodore Roosevelt. Zon reviewed models of community forestry in France, Switzerland, Germany, and Austria. In a later publication Zon further developed his analysis of communal forests in Europe by examining their origins and their links to the evolution of German city-states, feudal forms of land tenure, and formal town planning. Dana, after reviewing the roles of the German states in communal forest management, went on to advocate their establishment in the United States (McCullough 1995:115).

The enthusiasm of Fernow, Zon, and Dana for federal support of community forests was not shared by the first chief of the Forest Service, Gifford Pinchot. Although he had studied forestry in Germany and was well aware of European traditions of communal forestry, Pinchot did not embrace the idea of community forestry. His initial priorities concerned the establishment of regimes of scientific forest management on national forests, particularly those in the western United States. McCullough (1995:114) argues that Pinchot was also leery of the increased federal government intervention in local government that federal support for community forestry would entail. He suggests that the inveterate politician in Pinchot was concerned about the negative political repercussions of advocating such intrusion. Although this may have been true with regard to Pinchot's reluctance to intervene in local government, it did not moderate his enthusiasm for asserting direct federal control over private cutover forestlands when owners resisted the implementation of sustained yield forest planning and management (Dana and Fairfax 1956/1980:124).

The lack of formal federal government support for community forestry did not stop the municipal or town forest movement. It continued to evolve primarily through the support of regional, state, and local governments as well as private interests and organizations. States with the strongest town forest movements were Vermont, New Hampshire, Massachusetts, Connecticut, and Maine. Support for municipal forests, from both private and public sectors, grew from concerns about the negative effects of timber shortages on timber and wood products–dependent local economies, as well as growing interest in the recreational and aesthetic values of forests (McCullough 1995:116). State involvement and support for town forests developed simultaneously with the institution of state forester, the establishment of permanent forestry commissions, and increasing state capacity—technical, institutional, and financial—for constructive engagement with forestry issues on private forestlands and acquisition of private lands for state ownership and management.

Nonprofit forestry organizations were also instrumental in mobilizing financial and political support for town and municipal forests. Some of the most important early nonprofits included the Connecticut Forestry Association, the Massachusetts Forestry Association, the Society for the Protection of New Hampshire Forests, and the Forestry Association of Vermont. These groups promoted town forests and the appointment of town foresters and tree wardens, lobbied for legislative changes to give private landowners incentives to sustainably manage their forests, and advocated for government purchase of key forest tracts (McCullough 1995:120). As with earlier efforts to promote community forestry, these associations often drew on European examples of enduring community forests.

Community Forestry on Federal Public Lands: A Lost Opportunity

The views of Fernow, Zon, and Dana regarding the desirability of federal support for urban-based forms of community forestry were mirrored by an analogous debate within the Forest Service after 1910 regarding community forestry on the national forests. One of the primary advocates of community forestry on national forests was professional forester, regional planner, and political activist Benton MacKaye. Although he is better known for his proposal to create what was to become the Appalachian Trail (which in its current form bears little resemblance to the community-based regional development planning model of which the Appalachian Trail was originally only one part), Benton MacKaye's early career was with the Forest Service and later the Labor Department. MacKaye's ideas, although never actualized in policy by the Forest Service, are important for a variety of reasons. First, although he subscribed to many of the main tenets of the Progressive Era, including faith in science, the importance of efficiency, and the need to minimize waste, MacKaye was also highly critical of the dominant purposes to which Progressive Era tools and means were applied. MacKaye's counternarrative, as it were, opens up a space in the dominant Progressive Era discourse for entertaining alternative readings of the use of science, readings that are especially relevant for current attempts to strengthen the civic characteristics of the practice and culture of science. Second, MacKaye's ideas regarding the importance of addressing the social aspects of national forest management almost hauntingly presage the current concerns of the community-based forestry movement with regard to public lands management. In this respect, Benton MacKaye's ideas, although perhaps 60 years ahead of their time, also demonstrate the depth of the domestic roots of the community-based forestry movement in the United States.

MacKaye was a member of the first generation of professional foresters in America. After graduating from Harvard's forestry school in 1905 he conducted groundbreaking research on the effects of forest cover on stream flow in New Hampshire's White Mountains. His research, which correlated deforestation with irregular stream flow, provided the scientific basis necessary to link headwater forest management with the navigability of downstream waterways. This established federal government authority, based on the federal government's constitutional authority to regulate navigable waterways, to purchase private cutover lands granted in the 1911 Morrill Act for inclusion in the national forest system. A progressive himself, MacKaye was also a member of a small but influential group of intellectuals—economists, muckraking journalists, and writers—who resisted the potential elitism of preservationism and sought to push conservation politics in a more radical direction (Sutter 1999:555). After moving to Washington, D.C., in 1911, MacKaye joined the group of radical reformers known as the "Hell Raisers." During the next 10 years MacKaye developed plans for community-based settlement (he called it "colonization") of agricultural and forestlands that sought to integrate environmental and social concerns. The plans and the draft bills he authored were born of a critique of the graft, land speculation, and inequality of access to resources associated with the disposition of the public domain. In many respects MacKaye's emphasis on community-based development, democracy, equity, and cooperative institutions resonated with John Wesley Powell's blueprint for arid land development contained in his *Report on the Lands of the Arid Region* (Stegner 1953).

One of MacKaye's overarching interests was to unite the goals of the Forest Service and the Labor Department in an integrated plan that would guide federal land management actions toward sustainable resource development and sustainable labor regimes (Sutter 1999:556). MacKaye was acutely aware of the association between extractive resource regimes and exploitive labor relations. The social conditions associated with prevailing logging practices on private lands prompted Benton MacKaye's calls for the federal government to play an active role in fostering sustainable place-based communities. The dominant "cut and run" logging practices of the day depended on transient lumber camps peopled by single men who worked long hours for low wages—hardly conditions that promoted a rich community life and strong civic institutions. Indeed, the dangerous working conditions, inadequate wages, and long working hours in the woods and in the mills led to widespread strikes and labor unrest for both woodsworkers and millworkers in the early twentieth century, especially in the Pacific Northwest. In 1916, in Everett, Washington, labor organizing activities on the part of the International Workers of the World culminated in a violent confrontation between activists and vigilantes that left seven people dead and many more wounded. MacKaye's visit to Everett soon after the "Everett massacre" spurred him to formulate an alternative vision of the social effects of forest management. He proposed that national forests be part of a communitarian resettlement program that would provide an alternative socioeconomic system to the dominant industrial economy. However, the federal government in general and the Forest Service in particular veered away from the socialist ideals and community development models embodied in MacKaye's thinking and writing. Postwar anticommunist sentiment made the adoption of MacKaye's suggestion even more remote. Despite the clear link between exploitive labor relations and resource extraction, the Forest Service resisted MacKaye's attempts to integrate labor issues into forest management, arguing that his ideas for sustainable colonization of public lands went beyond the Forest Service's mission.

MacKaye's 1918 article "Some Social Aspects of Forest Management," published in the *Journal of Forestry* the same year that he left the Forest Service to join the Labor Department, is a clear articulation of the social problems associated with national forest management and MacKaye's ideas for their redress. MacKaye's proposals read almost like a manifesto for the contemporary community forestry movement. They stand as a trenchant critique of the lack of attention to social issues in forestry schools, which partially accounted (and still does, to a large extent) for the lack of attention to the human dimensions of national forest management. This created what MacKaye called the "problem of the lumberjack": the lack of community life and unstable employment associated with itinerant lumber camps (or their contemporary analogy, trailer parks in rural communities). MacKaye contrasts the substantial energy and resources devoted to advancing forestry for consumer and business interests during the Progressive Era with the neglect of the social conditions resulting from forest management. Concerned with the social effects of mineral, timber, and agricultural development, MacKaye saw the direct links between the "nation's transient lumber and mining camps and . . . clear-cut forests, barren fields and hillsides stripped of their mineral content" (Sutter 1999:560). MacKaye's goal was to transform itinerant lumber and mining camps into sustainable and thriving lumber and mining communities. He recognized the parallel constraints and discrimination that workers, whether in mines, forests, or farms, faced, and he recognized that effective policy and programmatic responses would have to cut across these sectors. He argued that stable communities based on forest management would not emerge until forest practices shift from "mining" timber to more sustainable forms of timber harvesting. He went on to outline a set of specific proposals for helping communities organize and maintain their integrity. He called on the federal government to create working circles, which included logging and sawmill operations that would foster family and community health. He argued that these efforts should include the development of local institutions for self-government, educational facilities, cooperatives, and other forms of public assistance. He espoused a broad definition of community health, much more encompassing than the narrow equation of community stability with timber industry employment used by the Forest Service for the next six decades. MacKaye also called for cooperation between national forest managers and adjacent industrial forestlands so that by integrating forest planning and management on public and private holdings, more sustainable patterns of employment and community health would result. He even went so far as to suggest cooperative control of national forestlands to develop integrated and sustainable working circles that would address the "social and labor aspects of forest management" (MacKaye 1918:213).

MacKaye's discussion of cooperative control of public forests resonates with current proposals advocating the creation of charter forests (despite their more limited focus). His emphasis on employment and labor issues, democratic process, and civic life clearly reflects the belief that people's franchise in the nation's public lands was based on citizenship as well as occupation and that these rights conditioned government forest rights based on territorial control. Consequently, forest workers had a right to sustainable work, good employment conditions, and participation in vibrant rural communities, and the federal government had a responsibility to provide for the exercise and fulfillment of such rights. For many forest-dependent communities, however, these rights were neither exercised nor fulfilled as, in the case of the timber resource, they were sold to the highest bidder. The challenges associated with the realization of forest rights based on occupation, in addition to those based on citizenship and ownership through due process and territorial control, respectively, constitute some of the biggest hurdles that the community forestry movement currently faces. MacKaye concluded his article by warning that if the "unsolved and menacing labor problems" (MacKaye 1918:213) and the social aspects of forest management were not resolved through government initiative, then the "problem of the lumberjack" would worsen. As Chapter 3 shows, MacKaye's warnings went unheeded, and what he foretold came to pass.

The ideas presented in his 1918 *Journal of Forestry* article were developed further in his report "Employment and Natural Resources," which he finished a year later while working for the Labor Department. This report proposed a two-pronged strategy for employment generation: short-term employment generation programs and long-term investment in the development of agricultural, mining, and forestry settlements. As he did in his 1918 article, MacKaye criticized government programs that provided short-term aid to individuals. Instead he advocated government assistance that developed community capacity and helped transform lumber and mining camps into lumber and mining communities. Within a context of chronic instability of labor demand, MacKaye proposed a variety of alternative employment opportunities that would restore direct working relations between people and natural resources, provide "wise resource stewardship," and "counter exploitative labor practices" (Sutter 1999:559). The report also contained a well-developed critique of the individualistic nature of the nation's homesteading policy, which provided opportunities for extensive land speculation and graft and did little or nothing to promote sustainable communities. MacKaye proposed an alternative homesteading blueprint, one in which individual land ownership was linked directly to use and was "subordinated to the long-term viability of the communities and of the resources themselves" (Sutter 1999:559). He believed that this quasipublic ownership would reform resource use.

"Employment and Natural Resources" contained a harsh critique of the effects of market forces on resource use and of the country's land settlement and resource development policies. However, it was also firmly rooted in Progressive Era values of efficiency, scientific expertise, and rational, centralized planning. This is interesting because although MacKaye was a progressive conservationist who espoused Progressive Era tools and methods, he deplored the ends to which other progressives applied those tools, that is, for economic and political purposes that he felt sacrificed social and environmental equity. MacKaye argued that the broader notion of social efficiency had been artificially narrowed to mean business efficiency. He consistently advocated community capacity building and argued that the role of the government was to "structure development in ways that facilitated democratic communities and guarded against the accumulation of power over labor and resources" (Sutter 1999:560).

Although MacKaye's writing and legislative proposals linking social, ecological, and institutional concerns never gained favor, his ideas compel a revision of our view of the relationship between the legacy of the Progressive Era and current attempts to promote community-based forestry on national forests. First, they illustrate that community-based forestry is not as new as many people think: Many of the dominant ideas were raised and advocated by MacKaye and others in the late nineteenth and early twentieth centuries. Second, MacKaye's politically liberal viewpoints and his emphasis on using the tools of the Progressive Era for community forms of development and resource management create an important space in the Progressive movement for entertaining and supporting the social and ecological objectives associated with the current community forestry movement. Progressive Era legacies, especially the role of science and the trained expert in defining and acting to achieve the public interest in an ostensibly value-free and politically neutral environment, are responsible for many of the contemporary institutional and policy elements with which the community forestry movement is currently grappling. MacKaye's ideas of socially responsible forest management illustrate that the Progressive Era's faith in rationality could be taken in different directions, including those that support and strengthen current community forestry initiatives. The Progressive Era was a moment in history that offered multiple models for forest conservation. The policy priorities and programmatic directions that were adopted to guide federal forest conservation and Forest Service policy, particularly with respect to social issues related to forest management, certainly were not the only ones that could have developed out of the Progressive Era. Understanding why policy priorities and programmatic directions were chosen that did not support MacKaye's goals of transforming lumber camps into sustainable communities based on principles of democratic participation, stability, justice, and equity (an issue taken up in Chapter 3) is central to understanding the depth of the challenge faced by the contemporary community forestry movement.

That the Forest Service did back away from MacKaye's ideas is symbolized by his transfer to the Labor Department and the fact that none of his suggestions for addressing the social dimensions of forest management were taken up as policy initiatives. Indeed, the Forest Service's rejection of community forestry and rejection of the connection between labor, communities, and resource management, which began under Gifford Pinchot's leadership, continued through until the 1930s when, at Franklin D. Roosevelt's request, the Forest Service created a community forestry program. Housed initially in the Division of State Cooperation in the Branch of State and Private Forestry Divisions, the program adopted the following definition of community forests:

> A community forest consists of lands owned and operated for forestry or allied purposes by a village, city, town, school district, township, county or other political subdivision, or by other community or group enterprises, such as schools, hospitals, churches, libraries, 4-H Clubs, Boy Scouts, Girl Scouts, and Camp Fire Girls. Locally a community forest may be known as the town, city, county, school, or municipal watershed forest; village or town or memorial woods; or community forest. (McCullough 1995:183)

From the start the program emphasized public education about the importance and potential of community forests. It worked primarily through state foresters and the Extension Service officials who themselves had direct contact with the committees, associations, and individuals involved in community forestry. The Forest Service produced and distributed a large number of pamphlets and publications designed to publicize community forestry, often drawing on examples of successful community forests in New

England. The primary emphasis of the community forestry program was on timber production in community forests; aesthetic, watershed, and recreational purposes were secondary (McCullough 1995:185). The program also maintained census information on the number, type, and acreage of community forests. By 1948, George Duthie, who was in charge of the community forestry program, estimated that nationwide there were more than 3,000 community forests with a total acreage of almost 4.5 million acres.

From the start, and notwithstanding its enthusiastic publications, the Forest Service's support for community forestry was tempered by skepticism about the ability of communities, municipalities, and organizations to successfully manage forests for long-term timber production. McCullough (1995:186) argues this is why the community forestry program never received legislative or financial support, and he suggests that doubts about the financial viability of community forests explain why a key opportunity to provide federal support for community forestry was not signed into law in 1941. The opportunity had arisen as a result of a joint committee on forestry established by Congress in 1938 to investigate the conditions of American forests. Of particular concern were the large tracts of low-productivity private forestlands. The Bankhead Committee recommended that federal funds be used to expand the number and area of community forests. However, Roosevelt, Secretary of Agriculture Wickard, and Acting Chief Forester Clapp, working with the Bureau of the Budget, reworked the committee's proposed legislation to eliminate federal support for community forestry, primarily out of concerns that federal subsidies for community forestry would not be "self-liquidating" and allied issues concerning the unpopularity of federal involvement in state and local government affairs (McCullough 1995:187). Although the Forest Service's community forestry program continued for another 8 years within the newly formed Section of State and Community Forests within the Division of Cooperative Forestry Management, the program effectively came to a halt with George Duthie's retirement in 1949. In the 3 or 4 years preceding Duthie's retirement there were no publications on community forestry, and the program consisted primarily of census activities.

Support for community forestry continued for a few more years within the Society of American Foresters (SAF) and the American Forestry Association (AFA). The SAF in 1941 had created a community forestry committee that espoused a definition of community forestry similar to that of the Forest Service. However, unlike the Forest Service's definition, the committee's definition privileged woodland conservation over timber production. The committee focused on two important aspects of community forests: their ability to provide employment and the conservation and development of forests for multiple uses. The committee produced annual reports on the status of community forests, often calling for improved forest management and strengthened state-level forest management capacities for doing so. When its attempts to convince the Forest Service to provide more financial and institutional support for community forestry withered with the demise of the Forest Service's community forestry program, the SAF turned to the AFA for support. However, the AFA did not respond enthusiastically. Although it did create a national committee on community forestry in 1952, it dissolved only a year later and it accomplished little other than to prepare a community forestry census published in the spring 1953 issue of *American Forests.* The SAF community forestry committee suffered a similar fate soon after; it was dropped from the SAF program in 1954. The disappearance of community forestry programs from both the Forest Service and national professional forestry associations indicates the extent to which the primary emphasis of forest management had veered away from the social concerns of early advocates of community forestry such as Bernard Fernow and Benton MacKaye.

Conclusion: Community Forestry, Equity, and Social Justice

The historical antecedents of community forestry discussed in this chapter are linked to issues of equity and social justice. These twin issues constitute an important common bond that joins these historical antecedents with the contemporary community forestry movement. Improving equity and achieving social justice for resource-dependent people and communities that have been disenfranchised from the forests on which they depend are central concerns and a continuing challenge of community forestry today.

In northern New Mexico, the cradle of a rich, centuries-old community forestry tradition, increasing restrictions on customary access to forest resources, culminating in the cancellation of wood-collecting permits in the Carson and Santa Fe National Forests in the late 1980s, severely affected Hispano communities. This increased already severe poverty levels and outmigration and reduced dwindling community capacity in the area. There is a tragic irony that today Hispano residents are living near popular and chic urban areas and on land worth hundreds of thousands of dollars, yet they are unable to secure access to subsistence resources that both sustain and define them. One of the primary factors motivating community forestry groups in northern New Mexico, such as Las Humanas Cooperative and La Montana de Truchas, is to address these social justice issues by providing forest-based employment and strengthening community capacity through forest ecosystem restoration and stewardship.

Native American traditions of community forestry, long suppressed by federal policies and programs that sought to eliminate, relocate, or assimilate indigenous people, are being gradually revitalized as American Indians claim their rights as sovereign entities, reassert control over their lands and territories, and strengthen their institutional and technical capacities for resource management both on and off reservations. Native American forms of community forestry, though only peripherally addressed in this book, are clearly and centrally linked to issues of equity and social justice. The emerging collaborative resource management agreements between American Indian groups, some public lands agencies, and, to a lesser extent, private forestland owners represent a radical departure from historical trends, one that carries with it the potential to empower and validate indigenous forms of community forestry and the knowledge it encodes.

The focus on equity and collective benefit also runs through the discussion of municipal forestry in New England. Then and now, town and municipal forests provide important collective public goods such as green space, income, and ecosystem services in urban environments. Occasionally, as in the case of poor farms in the eighteenth and nineteenth centuries, they directly benefited low-income groups and constituted part of a community's safety net. Today, they help anchor efforts to stem the tide of forest fragmentation and parcelization. Contemporary community forestry initiatives strive to maintain the economic viability of nonindustrial forestland ownerships in New England. They seek to counterbalance the economic advantages associated with the economies of scale and trade liberalization of our market system that benefit larger industrial forestland ownerships.

Finally, Benton MacKaye's concerns about working conditions on public and private forestlands and his adamant calls to give the social aspects of forestry as much attention as the economics and ecology of forestry clearly presage the focus within community-based forestry on community well-being and civic society, especially in relation to public forestland management. In particular, MacKaye's efforts to elevate the importance of forest worker issues within mainstream forestry are prescient given the current need in community forestry to embrace the social justice and equity concerns of all forest workers. Thus all four of the antecedents of community forestry discussed in this chapter contain within them important equity concerns and implications. They constitute a rich tapestry that informs and reinforces the importance of equity in the contemporary community forestry movement.

CHAPTER 3

Setting the Stage for Community Forestry

> . . . if we desire to have a system of forestry in this country which is concerned only with wood supply, streamflow, and their *material* byproducts, then we should pay no attention to the social aspects of forest management. But if we pursue this policy we must not be surprised in future times of crisis if the labor situation, in the industry which is ultimately our charge, becomes acute and grows worse instead of better; for fundamentally it will be "up to us" for our failure to prepare.
>
> —*MacKaye (1918:214)*

This chapter examines the social and ecological effects of the dominant trends of forest resource management in the twentieth century. This provides the basis for understanding the rise of community-based forestry as a social movement offering an alternative vision of forest management from that which dominated much of the last century. This vision contains many of the elements of late-nineteenth- and early-twentieth-century community-based forestry discussed in Chapter 2. By exploring how and why the dominant forest management paradigm of the twentieth century marginalized earlier forms of community-based forest management models, practices, and ideas, this chapter also identifies some of the current challenges faced by the community forestry movement.

Three core themes help explain why early community-centered forest management regimes, such as municipal or town forests in New England, communal range and forest management traditions among Hispanic communities of the Southwest, and Native American forest management practices, atrophied or declined throughout most of the twentieth century. They help explain why the more socially progressive strands of the Progressive movement, represented by Benton MacKaye, Bernard Fernow, and others, were not embraced by other Progressive Era foresters or institutionalized in Forest Service policies and programs or in state and land grant university forestry extension programs. These themes also provide a general framework for interpreting trends and patterns within private industrial forestland management. Although not attempting a detailed analysis of social and ecological change across the rural forested landscapes of this country, this chapter sets the broader context for understanding the rise of community forestry by examining the development of the conditions that led to its emergence. Furthermore, the analysis of the marginalization of the historical antecedents of community forestry identifies the forces opposing the current community forestry movement; thus this chapter lays the foundation for understanding the movement's rise and the barriers and obstacles it must overcome.

The three core themes are the consolidation of power and authority away from local communities within centralized public agencies and nonlocal interest groups at regional and national levels within a pluralist democracy and policy-making process, the high valuation of knowledge based on the dominant models of scientific research and the importance accorded to those who possess that knowledge (i.e., scientists and "experts"), and treating natural capital as income and allowing it and human capital to be converted to financial capital without regard to the ecological and social implications of doing so. This last point includes the consolidation of large industrial landholdings that occurred as a result of the interaction between policy and free market capitalist economics. Although these processes played out in different ways in different regions of the country and in different property right systems, they nevertheless provide a useful analytic for considering the dominant patterns of forest management and the genesis of community-based forestry.

The rest of this chapter provides a brief overview of how the core themes informed the forms, functions, and social and ecological outcomes of forest management on public lands, private industrial forestlands, and nonindustrial forestlands. The next section begins with a review of Progressive Era goals and values as they informed the institutional structure and early policies and programs of the Forest Service as well as forestry extension on private lands. The centralized, hierarchical, expert-driven Progressive Era models of forest management that were institutionalized at the beginning of this century were consistent with the then prevailing assumptions about science, and public administration and the role of government in many other sectors of society as well. Indeed, they informed and supported the rise of what has been called the American administrative state (Lee 1995). Thus, insofar as community-based forestry represents an alternative ordering of political and social relationships that challenges the dominant institutional paradigm, community-based forestry springs from the same set of critiques and strives for similar social objectives of other community-based movements such as urban redevelopment, brownfield reclamation, and environmental justice movements.

Progressive Era Principles for Forest Management on Public and Private Forestlands

The legacies of the Progressive Era—the disenfranchisement of rural communities from forest management policy and planning processes as a result of the rise of interest group politics and from the science of forest management through the discounting of local knowledge and the bias of science toward commodity extraction—have played key roles in creating the conditions that led to the emergence of community forestry, especially on public lands. Thus, it is worthwhile to briefly review some of the key elements of the narrow brand of progressivism (as opposed to the more populist versions espoused by MacKaye, Fernow, and others) that were institutionalized in the public land management agencies and the extension organizations and programs for forestry on private lands.

The Progressive Era was a period of unbridled enthusiasm for science and the ability of technically trained experts working from within scientifically organized and politically neutral bureaucracies to articulate and achieve the public good. Progressive leaders attempted to preserve the ideals of individualism enshrined in liberal democracy while simultaneously controlling the excesses of the increasingly powerful bureaucratically organized corporations (Williams and Matheny 1995). Most progressives, reflecting their somewhat elite backgrounds and biases, also did not want to share real political power with the lower classes. They generally were leery of and distrusted the ideals represented by lower-class movements such as the Grange and Populist movements of the end of the nineteenth century. Faith in science and in neutrally competent bureaucracies, both of which were important aspects of the new rationally organized society, was a common progressive response to the threat of corporate monopolies and to the specter of power sharing with the masses.

Progressive leaders emphasized the role of trained experts working within bureaucratic organizations in identifying the public interest and then taking the necessary steps to achieve it. The field of administrative science, also developed during this time period, was used by progressives to develop a science of rational administration to guide the structure and function of bureaucracies. The theory held that bureaucracies organized in accordance with rational administrative procedures would be impervious to political influence by elected politicians. Similarly, scientific expertise located in such bureaucracies would be able to develop policies that best achieved the public interest and would be unencumbered by democratic decision-making processes. Substituting scientific expertise for participatory democratic institutions for identifying and achieving the public interest is a hallmark of the Progressive Era. Williams and Matheny (1995) argue that the Progressive Era "transformed the discourse of American democracy by shifting discussion away from concern over participation in democratic politics to a focus upon neutral, scientific criteria for judging public policy" (p. 12).

People such as Gifford Pinchot embraced the principles of the Progressive Era with almost religious fervor. Gifford Pinchot distilled and interpreted Progressive Era ideas and used them as guiding principles for the agency he directed from 1898 to 1910. Unlike Fernow, Dana, and MacKaye, Pinchot did not embrace the populist strands of progressive thinking. Pinchot's progressivism was based on an elite view of science and social relations. He firmly believed that through the science of forestry, "the greatest good for the greatest number" would be achieved. And the practice of forestry itself required professionals trained in scientific forest management. He was as adamant that trained foresters knew what was best for the public as he was concerned about curbing monopolistic tendencies in the fast-growing capitalist economic system. This was the basis for Pinchot's reservations about the viability of community forestry in the United States; he believed that communities did not possess the necessary scientific expertise to sustainably and economically manage forestlands. Instead, public forests would be managed most efficiently by a skilled cadre of professional foresters in an organizational structure that insulated them from developing close social relations with communities to prevent those communities from influencing the forester in such a way as to shift forest management away from achieving the greatest good for the greatest number.

These progressive principles informed the reservation of national forests from the public domain in the late nineteenth and early twentieth centuries and the establishment of science-based forest management regimes in those areas. This pattern of territorial control and science-based management constituted a system of public forest ownership that slowed degradation of those lands while increasing the value of private industrial forestlands.[1] The Progressive Era interest in thwarting monopolistic corporate behavior (especially with respect to "public" resources) and in insulating public forests from "the masses" (in this case rural communities, grazers, and forest workers, especially people of color) was shared by progressive leaders such as

Pinchot, timber barons interested in restricting supply, and early conservationists such as John Muir (Romm 2002:125). The powerful combination of these elite forces was able to disenfranchise rural communities from public forests by delineating territorial boundaries and imposing scientific forest management regimes. As a result, in some areas, such as among northern New Mexico Hispano communities, previous patterns of community-based forest management were supplanted and gradually withered.[2]

Pinchot's commitment to establishing scientific forest management regimes did not stop with public lands. If anything, his zeal for implementing rational forest management was even greater when it came to private forestlands; forest degradation and the resulting threat of a timber famine were greater on private than on public lands. The extent of Pinchot's zeal is indicated in the following oft-quoted passage:

> The fight to conserve the forest resources of our public domain has been won. . . . Another and a far bigger fight has begun, with a far greater issue at stake. I use the word *fight* because I mean precisely that. . . . Since otherwise they will not do so, private owners of forestland must now be compelled to manage their properties in harmony with the public good. The field is cleared for action and the lines are plainly drawn. He who is not for forestry is against it. The choice lies between the convenience of the lumbermen and the public good. (In Dana and Fairfax 1956/1980:124)

Pinchot went so far as to advocate direct federal control over private forestlands if the owners did not implement a science-based forest management plan. Despite vigorous debate, Pinchot's support for direct federal intervention in private forestland management never won legislative approval. Instead, the U.S. legislature opted for state regulation of private forestlands and passed a series of laws (Clark–McNary, 1924; McSweeney–McNary, 1928; and Knutson–Vandenberg, 1930) designed to cooperatively build a variety of state capacities for effective intervention in private forest management. These included cooperative federal–state programs for firefighting, reforestation and management, and research.

From the perspective of community-based forestry, the decision by the federal government to work collaboratively with states on forestry issues rather than to intervene directly has two important implications. First, it initiated the important process of developing institutional capacities, policies, and programs within state forest departments and the Forest Service (e.g., the Division of State and Private Forestry) for working directly with private landowners and rural communities and with county governments. State and county foresters, and some county governments, have been at the forefront of recent innovations in community-based forestry; in many cases the institutional origins of these state and county capacities derive from this legislation. State and Private Forestry in the Forest Service, for years marginalized within the agency, has over the last decade emerged as a leader in supporting community-based work and community and worker engagement in land management. Second, the primary thrust of the early federal–state cooperative programs and the forestry extension efforts that emerged from them have historically emphasized the application of silvicultural principles of scientific forestry for the purposes of maximizing timber growth. Thus, extension, education, and outreach for private landowners reflected the biases of Germanic forestry and, for the most part, ignored the diverse needs and concerns of private forestland owners, especially nonindustrial forestland owners.

The inability of forestry extension to adequately address the needs of nonindustrial private forestland owners in terms of information and financial assistance stems from a variety of factors, many of which relate to the various diseconomies of scale faced by nonindustrial landowners and their diverse land management objectives. Landowners with small to medium holdings face significant difficulties with respect to issues such as market access, information regarding timber and other forest product values, access to loans and other sources of capital, and developing value-added products. Many nonindustrial private forestland owners also practice a mix of forestry, agriculture, and animal husbandry and thus have a diverse set of land management objectives. Forestry extension programs, geared more for large-scale industrial, forestry-only operations, were ill-prepared to respond to the information needs of landowners who integrate forestry with other land management activities. Extension foresters, trained in the progressive model of scientific forest management and focused on timber production, also have historically gravitated toward industrial operations because of similar views concerning the purposes of forest management and the economies of scale presented by their size. Also, in some regions, particularly the South, the provision of credit and subsidies, access to beneficial government programs, and technology and information transfer have been withheld from African American and limited-resource landowners because of institutionalized racism and class bias. Although in recent years this phenomenon has been widely publicized in the context of agriculture (witness the $3.5-billion 1996 class action lawsuit against the U.S. Department of Agriculture brought by African American farmers), similar dynamics have obtained in forestry.[3]

Although Pinchot's extreme views regarding federal intervention in private forestry were not enshrined in policy or legislation, they nevertheless illustrate the willingness of progressives to sacrifice due process and participatory democracy in the interests of scientific management for achieving their definition of the public interest. It suggests that the pursuit and achievement of the public interest, as defined by scientific expertise, could justify authoritarian and undemocratic modes of governance. The firm belief that there was one best policy discoverable by scientific experts and that due process, participation, and democratic values were secondary to the task of achieving the public interest was a common theme of Progressive Era thinkers, especially with regard to government regulatory policy making (Williams and Matheny 1995:13). Furthermore, because issues often were defined in purely technical rather than political terms, there was little basis for resolving fundamental value conflicts: The technically driven notion of the public interest drove out alternative notions of the public good derived from other value systems. The centrality of scientific expertise in natural resource management—the belief that there was one discoverable public interest and that its discovery was a technical rather than a political matter—constituted a Progressive Era legacy that was not seriously challenged until the 1970s.

The insulation of scientific expertise from rural communities, through the placement of technically proficient professionals within hierarchical organizations, virtually eliminated the possibility of rural communities or workers sharing their knowledge or having their desires and interests regarding public and private forest management heard. Progressive Era class biases and ongoing institutional racism, combined with the willingness to dispense with participatory democracy to achieve a predetermined goal, guaranteed its elimination, especially with regard to minority communities. This issue, the muzzling of rural communities' voice with respect to forest management issues and the distribution of the costs and benefits of forest management, later became one of the more important concerns of the community forestry movement.

The New Deal's Response to Community Instability and Resource Degradation

The Depression era social and environmental problems that confronted society and government provided New Deal leaders opportunities for innovative policies and programs based on the ideas of progressive thinkers such as MacKaye. High levels of unemployment coupled with continued labor unrest in much of the Pacific Northwest suggested that the social and labor aspects of forest management about which MacKaye had warned had only grown worse. The specter of timber shortages on industrial forestlands and concerns about soil erosion and rangeland degradation on public lands indicated the widespread nature of environmental concerns. Concomitant with these issues was an unprecedented faith in the ability of government and science to develop solutions to complex social and environmental problems.

Witness, for example, the optimism with which integrated river basin management efforts such as the Tennessee Valley Authority were greeted (and exported abroad) and the expectations that such efforts would usher in an era of grassroots democratic fulfillment and science-based resource management. These efforts reflected the depth of faith in science, government, and bureaucracy that the Progressive Era had bequeathed to the next generation, as well as the general consensus that it was the government's legitimate role to take the lead in addressing such complex issues. Here, it seems, was an opportunity to address in an integrated fashion the twin issues of how to develop environmentally and socially sustainable resource management.

However, actual accomplishments fell short of the mark. A review of what was and was not achieved provides useful context for the contemporary community forestry movement and highlights the challenges it faces, including those that can be traced back to this period. The primary government response to the "problem of the lumberjack" consisted of the various job creation programs for which the New Deal is known. Though providing short-term employment and accomplishing many conservation-related tasks, this approach undermined work and occupation as a legitimate basis for people's forest enfranchisement. Short-term employment generation programs were only temporary salve for systemic worker issues. Not only did the resource agencies not address the underlying structural issues about which MacKaye had written so forcefully, but from a policy standpoint they also contributed to the severance of labor issues from the environmental issues with which they were so intimately connected. Simultaneous debates about how to slow rangeland degradation and timber harvesting on industrial lands eventually culminated in the 1934 Taylor Grazing Act and the 1944 Sustained Yield Forest Management Act. Both the grazing districts and the sustained yield units that these acts authorized were based on principles of rights of use, access, and benefit rooted in place-based notions of forest enfranchisement. They both excluded non–place-based rights holders, further weakening the forest claims of this diverse group. The result was an unfortunate narrowing of the basis for legitimate forest enfranchisement from that which MacKaye had espoused. Rather than the broader notion of enfranchisement based on occupation, place, or citizenship that MacKaye had articulated, the New Deal's approach to the social and environmental issues of the time segregated labor issues from the broader discourse of community development and substantially weakened the forest claims of non–place-based groups, especially people of color; enfranchisement based on place (and later citizenship) was strengthened while enfranchisement based on occupation diminished.[4]

The Forest Service's involvement in community development in the interwar years consisted primarily of implementing New Deal employment generation programs. These programs were designed to alleviate short-term economic stresses associated with high unemployment. The Civilian Conservation Corps, created by the Emergency Conservation Act of 1933, and similar forest conservation and employment generation programs created by the Federal Emergency Relief Act of 1933, the National Industrial Recovery Act of 1933, and the Works Relief Act of 1935 provided employment to large numbers of men otherwise unable to find jobs. These programs, which were almost entirely restricted to public lands, subsidized a wide variety of conservation activities including reforestation, forest protection and improvement, soil conservation, and recreational development. And while they lasted, they provided a means for men (and some women) to support themselves and their families. However, they were not part of a broader programmatic effort to stabilize and strengthen rural communities, and they did not address the underlying structural inequalities that gave rise to the widespread labor unrest in forest-dependent communities in the Pacific Northwest in this period.[5] Issues of employment and community stability, as they related to forest management on private lands, were addressed through the concept of sustained yield.

Proposals calling for the implementation of sustained yield forestry to reform the historical "cut out and get out" practices of private timber industry came from both industry and the Forest Service in the late 1920s and early 1930s (Robbins 1989:14–15). However, without exception, these proposals were shorn of the important social agendas that had been the centerpiece of MacKaye's programs and initiatives. In response to the financial losses, overproduction, and rapid forest depletion that characterized the timber industry in the 1920s and 1930s, especially in the Pacific Northwest, industry organizations called for cooperative agreements between industry and the government to control overproduction and stabilize the timber industry.[6] The principle adopted to achieve this goal was sustained yield, the hallowed principle of Germanic forestry that dictates that cut shall not exceed growth. From the industry perspective, sustained yield meant curbing the overharvesting of timber and, by restricting supplies in a coordinated fashion on both public and private commercial timberlands, increasing market prices for forest products to create stable and profitable business conditions. The primary beneficiaries of the proposals would be the large firms that would enter into cooperative agreements with the Forest Service. Smaller operations, excluded from these arrangements, opposed this version of sustained yield forest management in the belief that the production restrictions discriminated against them (Robbins 1989:15). Although originally designed to control overproduction and stabilize market prices, some industry representatives were attuned to the political importance of social issues and later added community stability as an element and goal of sustained yield (Mason 1927:625).

During this same time the Forest Service also supported proposals for achieving sustained yield forest management. However, unlike industry representatives who were primarily concerned with industry stabilization, progressives within the Forest Service such as Ferdinand Silcox, who became chief of the Forest Service in 1934, and former chief Henry Graves were concerned about the social context of forest management and, in particular, the health of timber-dependent communities and local economies (Robbins 1989:15). Silcox and Graves were extremely critical of industry's cut-and-run timber harvesting practices, especially in terms of the resulting social dislocation and the lack of attention to reforestation and allied forest investments.[7] Although Forest Service and industry interest in achieving sustained yield forest management shared a focus on curbing overproduction through coordinated reductions in timber harvest levels, only the Forest Service version of sustained yield incorporated commitments to community stability. And, as was true for the next half century, community stability, to the extent that it was defined, was conceived of in the narrow context of timber harvest as it related to employment.

The debates over sustained yield, especially its role in stabilizing the timber industry, eventually crystallized in 1944 in the Sustained Yield Forest Management Act. The act authorized the creation of cooperative sustained yield units on public or public and private timberlands, and it guaranteed stable log flows for timber harvesting and processing firms within the boundaries of the unit. In most respects, the act, for which the Forest Service did not actively lobby, was modeled after the sustained yield proposals that industry representative David Mason had been developing since the late 1920s. Its passage by Congress reflected the political influence of the timber industry; the views of rural communities and small industrial and nonindustrial forestland owners were not represented during the legislative deliberations (Clary 1987:4). The earlier differences between industry and the Forest Service with respect to the extent to which sustained yield included social concerns and maintenance of community stability resurfaced in discussions about the act and the policies that would be based on it. The Forest Service emphasized the social objectives of the act, whereas "forest products executives . . . rejected the direct social implications of the sustained yield law" and instead focused on the industry stabilization aspects of the legislation (Robbins 1989:17). By sidestepping the social components of the act, the timber industry rejected an opportunity to contribute to the development of a long-term, mutually beneficial relationship between its own sustainability and local community well-being.[8]

Forest Service response to the act was less than enthusiastic. In both Washington, D.C., and the regional offices, Forest Service officials were reluctant to allow the timber industry monopolistic access to public timber, and they doubted industry's commitment to community stability (Clary

1987:5–6). Organized labor, small logging contractors, and communities outside unit boundaries also opposed the program and its place-based system of preferences. These groups resented the preferential treatment afforded the large companies within sustained yield units that received guaranteed access to federal timber. Only one cooperative sustained yield unit was eventually established: the Shelton, Washington, unit involving Simpson Logging Company. Only five sustained yield units on national forestlands were created. Several units were short-lived and were discontinued because of criticism of their unequal distributive effects. Even the sustained yield units on national forestlands met with only limited success, especially with regard to the goals of community stability. For example, Krahl and Henderson (1998) show how the Vallecitos Sustained Yield Unit on the Carson National Forest in northern New Mexico has been almost continuously plagued by social conflict. The unit was developed to co-opt local opposition to reductions in grazing and firewood cutting allotments for Hispano communities by providing compensatory employment opportunities on the national forest (Clary 1987:9). Ongoing conflict between the designated operator and local communities, conflict between the Forest Service and the beneficiary communities, and opposition to the unit by communities located just outside the boundary of the unit have severely hampered the unit's performance and resulted in numerous calls to disband it. In her study of the Lakeview, Oregon, and Big Valley, California, sustained yield units, Cheek (1996) found that sustained yield units have the potential to contribute to community well-being but that their actual effects are a result of the interaction between the unit itself—preexisting internal baseline conditions such as physical, structural, industrial, and human conditions—and external factors, including changes in land management and the wood product industry. However, even where units have a positive impact on the community, Cheek argues that they do not create social capital (1996:114). Regarding the Flagstaff Sustained Yield Unit, a Forest Service report in 1956 concluded that the unit had not had any effect on "community developments or expansion" since its inception in 1949 (Clary 1987:15).

The Sustained Yield Forest Management Act foundered on the negative distributive consequences of its most important design principle, that of guaranteeing a supply of timber to timber operators and mills located within the sustained yield unit. It ignored that purchasers and forest workers have long operated in circles beyond "local," a fact that is even more true today with the mobile and largely Hispanic workforce. And it relied on business, which secured virtually exclusive access to timber, to equitably distribute local benefit. As a result, the communities, workers, and contractors located outside the unit and therefore excluded from the unit's benefits (and indeed, in some cases workers within the unit) opposed this preferential treatment.

The set of place-based preferential privileges enshrined in the act were experienced and opposed by those outside the units as a set of unfair, anticompetitive exclusionary practices. The excluded parties ultimately were able to prevent the establishment of new sustained yield units and contributed to the demise of the program.

The exclusionary processes within the sustained yield units are also evident in the efforts to avoid a "tragedy of the commons" on the open access rangelands managed by the Bureau of Land Management (BLM). To reduce rangeland degradation from overgrazing, the 1934 Taylor Grazing Act authorized the creation of grazing districts and instituted a system of grazing permits to limit the number of livestock on the range. Based on the principle of home rule, advisory boards made up of local place-based ranchers were established to implement the permitting system and give the BLM advice about range management and restoration. The criteria for determining who had priority for obtaining grazing permits that these advisory boards adopted excluded non–place-based migratory herders (such as the Basque shepherds John Muir disparagingly called "tramp sheepmen") and smaller livestock operations, especially those that had had to sell their ranches during the Depression. The advisory boards thus restricted access to public rangelands to families with large base landholdings and in the process excluded smaller operations and migratory people who had previously grazed their herds on these same rangelands and had developed customary rights to continue to do so.

Both the sustained yield units and the grazing districts privileged place-based actors and excluded non–place-based people from their benefits. In the case of the sustained yield units, the excluded parties had the political capital and the free market ideology to stop the expansion of the program. However, the non-Anglo migratory herders and small cattle ranchers, who were excluded from the grazing districts as a result of the exercise of the principle of home rule in prioritizing access to the range, did not have the political clout to challenge the abrogation of their grazing rights. As a result, the place-based system of prioritizing access to the range continues unchanged today despite many attempts to reform it. Both sustained yield units and BLM grazing districts were created to strengthen the stewardship capacities of place-based entities; in both cases representatives of those place-based interests who stood to gain from the units or districts helped draft the enabling legislation and implement their respective programs. And in both cases attempts to promote stewardship and community well-being within the units or districts depended on processes of exclusion that prevented others from accessing resources.

New Deal–era efforts to stabilize rural communities and foster sustainable resource management, such as the sustained yield units and BLM grazing districts, illustrate the difficulties of encouraging resource stewardship and social benefit for one group without excluding or harming the interests of others. In both cases prioritizing the rights of place-based groups disadvantaged and excluded non–place-based groups. A key challenge, then and now, is to develop inclusive decision-making processes that recognize and allow for the mutual accommodation of the legitimate claims of all groups, whether those claims are based on place, work, or citizenship. Recognizing the validity of forest enfranchisement based on work and occupation has been rendered more difficult by the separation of labor issues from mainstream community development theory that occurred in this period. The challenge of reintegrating labor and worker issues into forest management principles and practices has been made more difficult by the ways in which the Forest Service interpreted community stability and community well-being in subsequent decades.

Post–World War II Trends

This section reviews some of the dominant trends in forest management in the last several decades, especially with regard to the effects of these trends on rural resource dependent communities. As will be seen, community stability and community well-being, terms embraced by the Forest Service and used to justify collaboration between public and private forestland management, were defined narrowly to refer only to employment levels. Active community participation in public or private forest management planning and practice, other than through the provision of skilled labor for woods or millwork, did not exist. Additionally, public lands management agencies and private industrial forestland owners suppressed traditions of community-based resource management that were inconsistent with textbook forest management. This includes the forest grazing practices of Hispano communities in the Southwest on what had originally been communal land grants deeded to them by the Spanish government and local traditions in the Southeast of using fire to promote forage for livestock and to maintain the longleaf pine ecosystem. Additionally, key pieces of federal legislation in the 1970s that reflected the rise of environmentalism included processes for increased public participation in resource management, but the structure of those processes mitigated against the meaningful participation of workers, minorities, and many rural communities. Once again, people whose forest claims were based on occupation were excluded from decision-making processes that concerned the resources on which they depended.

Increased demand for forest products associated with World War II and concomitant labor shortages alleviated some of the previous problems related to overproduction, excess mill capacity, glutted markets, and high forest sector unemployment that characterized the interwar years. The post–World War II building boom and the consequent rise in demand for forest products further alleviated these problems. During this period of high demand, support waned for sustained yield policies as a means to curb production and stabilize prices, and concerns about community stability were similarly set aside as rates of timber harvesting (especially on public lands), capital investment, and forest sector employment increased. Dramatic increases in timber harvesting on public lands compensated for declining timber supplies on private industrial forestlands, which had decreased because of decades of harvesting beyond short-term sustainable levels. Between 1952 and 1959 the allowable cut on national forests increased from 4.38 to 10.6 billion board feet, and harvest levels had increased from 2.86 to 8.3 billion board feet. By the 1960s, harvest levels had increased to 12 billion board feet (Yaffee 1994:4). The timber supplied to western forest product industries by western national forests doubled from 15 percent in the early 1950s to 30 percent by 1962 (Parry et al. 1989:27).

By substituting the liquidation of national forest natural capital for the already depleted natural capital of industrial forestlands, rural timber-related employment levels were (temporarily) maintained, but the broader meanings of community stability and community well-being were never embraced. Defining community stability in terms of employment in the timber industry (as "a by-product of industry prosperity"), shorn of the broader connotations that Fernow, MacKaye, Dana, and others embraced, enabled the Forest Service to cloak its proindustry timber bias in the mantle of public and community service (Fortmann et al. 1989:44). Increased harvest levels in national forests, beginning in the 1950s, had more to do with making up the shortfall in private timber supplies and meeting demand than achieving broadly defined forms of community stability.

Throughout this period the Forest Service, as well as other federal agencies involved with forest management such as the BLM and the Bureau of Indian Affairs (BIA), continued the Progressive Era tradition of scientific forest management by trained foresters who assumed that the purpose of forestry was to maximize timber production. In this context the local knowledge of rural communities was discounted as subjective and unscientific, and alternative ways of construing the purpose of forestry that did not give primacy to timber production were considered illegitimate. Therefore, Forest Service employees continued to attempt to curb traditional grazing practices of Hispano communities in the Southwest, which they believed detrimental to forest health, and the BIA, in its management of forest resources on tribal reservation lands, gave little credence to Native American culturally rooted traditions of forest use and management. Rural forest-dependent communities and their knowledge were excluded from forest management planning and practices on public, private, and reservation forestlands. The primary point of engagement between rural communities and forest management was the use of their skilled labor for extractive forest product industries.

Throughout the 1970s and extending into the 1980s, a series of factors led to dramatic reductions in timber harvest levels on public and, to a lesser extent, private forestlands.[9] During this same period the timber industry was undergoing dramatic changes: Processing facilities were automated, raw log exports were increased, and an increasingly competitive global wood product market was developed. All these factors contributed to increased competition in the domestic industry and, at least in the western United States, to the "overcapacity" of mills. Reduced harvests and increased competition resulted in mill closures, started a trend of declining wages for workers, and led to dramatic reductions in wood product industry employment. These trends were exacerbated by the recession of the early 1980s.

The loss of jobs in the woods and the associated widespread mill closures had devastating economic and social effects on rural timber-dependent communities (see Lee et al. 1989). Because community stability had been long identified only with employment, as opposed to broader notions of community capacity, rural communities had few resources with which to manage the hardships associated with the drastic reductions in employment levels in the forest sector. The effects on rural communities in terms of increasing poverty levels and other indicators of social anomie revealed the hollowness of previous commitments to community stability and the precariousness of basing stability on timber harvest levels (Kusel 1991; Marchak 1990). It also laid bare the historical processes through which the natural capital of first nature, consisting of the centuries-old accumulation of biomass and energy in old-growth forests (Cronon 1991), had been liquidated and siphoned out of the timber-producing regions to enrich a small number of timber industry executives and leave in its wake impoverished forests and communities.

Although several factors were responsible for declining employment in the forest sector, the rise in power and influence of the domestic environmental movement played a central role in challenging the commodity and market orientation of forest management. In the late 1960s large segments of the (primarily urban) American public grew increasingly disenchanted with the Progressive Era models of natural resource management that prioritized commodity production and resisted calls for greater emphasis on noncommodity natural resources and more attention to the environmental consequences of resource extraction. Using a model of forest enfranchisement based on citizenship and the same model of interest group politics that industry representatives had earlier adroitly used, the rapidly growing environmental movement flexed its new muscles to lobby for key federal environmental laws. The legislative result was the passage of several key pieces of federal legislation in the 1970s, including the Endangered Species Act,

National Environmental Policy Act, National Forest Management Act, and Federal Land Policy and Management Act. At the state level, environmental protection laws modeled after the federal legislation were also passed. Several features of these laws, such as the mandate that public land planning teams be interdisciplinary, increased congressional oversight of agency activities, and unprecedented mechanisms for public participation in planning and management decisions, were direct challenges to the Progressive Era model of efficient, scientific management by trained experts insulated by their bureaucracy from "undue" influence from the public and politicians.

Two factors mitigated against the possibility of increased participation by rural people, both place-based and non–place-based, in managing the resources on which their livelihoods depended. The first was the continuation of policy making and political processes firmly rooted in the dominant model of pluralist democracy. This political framework, in which policy and planning directions are driven by interest group politics, favors groups able to mobilize resources and influence at state and national levels. These groups coalesce around specific issues and through electoral and other avenues influence policy making. In the context of forest issues, the key national interest groups were extractive industries and national environmental organizations. Throughout the 1970s and 1980s most rural resource-dependent communities did not have the capacity to effectively articulate their own concerns in state or national policy-making arenas. Meanwhile, beginning in the mid-1980s, court-ordered reductions in timber harvesting on public lands resulted in drastic reductions in timber harvest levels as well as legislative, policy, and planning gridlock. Perhaps the most dramatic example of this is the 1991 ruling of Seattle District Court Judge William Dwyer that halted timber harvesting in spotted owl habitat on public lands in the Pacific Northwest based on Forest Service and Fish and Wildlife Service violations of the Endangered Species Act and the National Forest Management Act.[10] In the West, what some rural communities perceived as the resulting "lock up" of natural resources led to a series of "sagebrush rebellions" and "home rule movements" against federal land management agencies and national environmental laws that were perceived to be unjustly imposed by nonlocal interest groups. Through policy-making processes that most rural communities had not been able to access, they had been excluded from the forests they had previously worked and relied on. As discussed in Chapter 6, proponents of community forestry in the western states have responded to this situation by seeking to create a political space separate from the traditional interest group politics of pluralistic democracy in which to foster more participatory and "republican" forms of democracy.

The second factor that mitigated against meaningful engagement of rural communities in resource planning on public lands concerns the ways in which agencies insulated their decision-making authority and monopolistic claims to management expertise from the public participation mechanisms mandated in environmental legislation. Agencies choose modes of public participation that limit genuine public involvement and exchange of ideas. Once public input has been gathered, "neutrally" competent technical experts within the agency make the final determination as to which plan option will be pursued (Moote and McClaran 1997:474). Public involvement in most public participation forums is actually quite limited. Though free to use a wide variety of participatory forms of public involvement, agencies almost always limit participation to the agency-controlled public hearing and the formal comment periods the regulations require (Moote and McClaran 1997:474).[11] The final determination regarding which option to pursue may be influenced more by political considerations and the search for decisions that will be legally defensible than by the desire to develop a plan that reflects an evolving and shared set of values and objectives and attempts to integrate a broad spectrum of concerns and interests.

Criticisms of this form of public participation derive from a more basic critique of the democratic institutions that undergird them. This critique has at least four elements. First, land use plans that result from this process are not based on the needs, concerns, and values of the people the plans affect. This is because they tend to be driven by rational scientific definitions of the public interest and how to manage for it. Second, the modalities of eliciting public participation in the planning process tend to favor well-organized interest groups and the people they hire to give "expert testimony." This can lead to interest groups and their experts dominating participation forums, with little or no input from the broader local (and nonlocal) public. The limited human and financial capital in rural areas can limit the ability of rural communities (and especially non–place-based groups such as migrant workers) to meaningfully participate in these forums, in terms of the skills needed to effectively participate, the time needed for such participation, and the resources necessary to make well-prepared, documented, and supported presentations. The rigid process and structure of these forums can easily lead to class-, race-, and ethnicity-based biases in public hearings. Third, the noninteractive nature of the participatory process mitigates against the development of trust between agency personnel and affected publics, prevents the development of shared values and goals for resource management, and makes it easier for agencies to ignore public input (Moote and McClaran 1997:475). As Kemmis (1996) points out, very little if any hearing actually occurs at public hearings. The one-time, one-way nature of most public participation forums does not encourage long-term involvement, exchange, and interaction between agencies and the public in which, based on collaborative research and monitoring, changes in a plan might be jointly developed and implemented. Finally, the traditional forms of public participation insulate agency decision-making discretion from public input. The process through which comments are analyzed and weighted and their actual influence on the planning process often are murky at best. The comment process itself is designed to meet the letter of the law regarding public participation without seriously engaging groups and people external to the organization. As a result, organized interest groups and the general public have little vested interest in or sense of ownership of the plans and management outcomes that result from such a process, primarily because the structure of the participation process preserves agency hegemony. This way of structuring public involvement does not provide opportunities for different interest groups to engage in dialogue together. It focuses on information gathering, not joint learning and decision making; it does not provide incentives for developing a civil political discourse about contentious and value-laden resource management issues; and it sets up the public agency as a final arbiter, responsible for weighing the different inputs it receives and deciding which planning and management direction to take (Cortner and Moote 1999:19).

The unorganized interests of rural communities and workers have not benefited from the participatory provisions of national environmental legislation. Because of the continued dominance of interest group politics, only the well-funded and well-connected special interest groups were able to effectively engage in natural resource planning. Rural communities realized that although timber industry lobbying groups and national environmental groups might have argued over resource-dependent communities, neither argued for them. In fact, national environmental organizations have had to struggle with the elitist legacy of their history and the resulting antirural and urban worker biases within their organizations, which, until recently, made it difficult for them to embrace worker and social justice issues.

Setting the Stage for the Rise of Community Forestry

By the end of the 1980s the factors discussed in this chapter created a set of conditions that was ripe for challenge. Long-term disinvestment in rural ecosystems and rural communities was a common feature across the country. Political and legislative challenges to the dominance of Progressive Era models of science, forestry, and administration had been mounted. However, because the challenges were still firmly rooted in the dominant model of interest group politics, they merely substituted one set of top-down policy approaches for another; the populist ideals of MacKaye, Fernow, and others were still in the wings, if not offstage altogether. Rural communities continued to be marginalized politically. Minority communities and workers were even more disfranchised than impoverished Anglo communities, although both shared disproportionately high rates of unemployment and other social maladies. In industrial forest management, the increasingly competitive global market for forest products, junk bond–leveraged corporate takeovers, and the rush to satisfy short-term investor demands for financial returns on their investments fueled the race to the bottom and resulted in widespread mill closures, outsourcing, and in some cases unsustainably high timber harvest levels to pay off high-interest junk bond debt.

The management objectives of many nonindustrial forestland owners, often cash-poor themselves, were often frustrated or threatened by one or a combination of the following factors: forest sector intermediaries who unfairly profited from incomplete landowner market information, limited or poor access to credit and other sources of capital, poor integration with markets for forest products and little or no opportunity for value-added processing, the threat of subdivision resulting from financial pressures caused by land fragmentation and increasing tax burdens, "high grading" of NIPF (non-industrial private forestlands) woodlots by unscrupulous contractors, financially burdensome forest practice regulations designed primarily for industrial operators in a climate of mistrust, and continuation of the historical legacy of inadequate or inappropriate forms of landowner assistance that did not reflect the landowner's desired integration of forestry with agriculture and animal husbandry and was embedded in structures of institutional racism. The dominant characteristic of forest management on public lands was gridlock. The stage was set for the emergence of a different way of relating communities and forests. Community-based forestry, though assuming different forms in different regions and taking root more rapidly in some areas than in others, is a coherent grassroots challenge to the dominant paradigm of forest management. The emergence of community-based forestry, discussed in Chapter 4, also signals a revitalization of its own historical antecedents that had lain dormant for most of the twentieth century.

CHAPTER 4

An Emerging Social Movement

> A social movement is a set of opinions and beliefs in a population which represents preferences for changing some elements of the social structure and/or reward distribution of a society. . . . A social movement organization is a complex, or formal, organization which identifies its goals with the preferences of a social movement or a countermovement and attempts to implement those goals.
>
> —*McCarthy and Zald (1977:1217–1218)*

> A social movement consists of a sustained challenge to powerholders in the name of a population living under the jurisdiction of those powerholders by means of repeated public displays of that population's numbers, commitment, unity, and worthiness.
>
> —*Tilly (1994:7)*

In February 1996, the Seventh American Forest Congress convened in Washington, D.C. As the name implied, this congress followed the first, held in 1882, and the six that followed roughly every 15 to 20 years thereafter. Congresses were held when forestry leaders felt public values and the policy and management sands were shifting and the need for dialogue was great. Unlike previous congresses that involved scientists and management experts exclusively, the Seventh American Congress convened more than 1,200 participants from a wide variety of backgrounds and with diverse experiences. Participants included wood products industry workers and executives, federal agency representatives, environmentalists, agency and university scientists, urban foresters, students, and others from rural and urban areas around the country. The Seventh Congress was billed as the people's congress, and despite high hopes among some for wood products industry–environmentalist rapprochement and for agreement on a future forestry research agenda that were not achieved, the dialogue among participants was, for the most part, extraordinary. In addition to the unprecedented dialogue, the congress process of engagement and the conversations therein made clear there was broad support—even demands—for more open and inclusive resource management decision making and more participatory resource management science. Also significant were the voices calling for a resource management to be more responsive to community forestry ideas. For those who had any doubts, the Seventh American Forest Congress made clear that the public was frustrated and no longer accepting business as usual and, equally important, a wide variety of people rallied around neither of the traditional poles of environment and industry as they called for new approaches and new processes for resource management.[1]

In rejecting traditional command-and-control approaches to resource management and resource science and supporting more open and community responsive resource management, these Seventh American Forest Congress participants reflected ascendant community forestry ideas. It is testimony to the power of the ideas that 6 years after the Seventh American Forest Congress, the only active group remaining from this not well understood but important event is the Communities Committee.

In this chapter we explore the conditions that gave rise to community forestry, the common themes that underlie its diverse regional forms, and some of the national and regional social movement organizations associated with community forestry. We argue that, consistent with sociological definitions of social movements, community forestry is a movement, with practitioners and advocates seeking changes in how people and resources are valued and changes in the social and institutional structures associated with resource management. Similar to some of the issues raised by participants of the Seventh American Forest Congress, specific examples of these changes include increasing participation in natural resource management decision making by communities and workers; modifying and strengthening institutional capacities for civic science and adaptive management; reversing the long-standing practice of capital extraction with insufficient reinvestment in the forest and resource-dependent communities and workers; developing investment and value-added production and marketing strategies that return value to forest ecosystems, communities, and workers; and expanding the basis for forest enfranchisement to include work and occupation in addition to place- and citizenship-based forms of forest enfranchisement. Almost without exception, community forestry consists of a populist challenge, rooted in participatory democratic process, to the narrow Progressive Era definition of the purpose of forestry and the narrow set of interests forestry has benefited for most of the last century. For many it represents a challenge to these narrow and dominant interests, including those that advance the model of scientific industrial-scale forest management. This challenge is as much about the failure of these interests to maintain diverse ecological structures and diverse forest habitat as it is about their failure to respond to local social and economic concerns.

To help advance these and other forms of social change, including reinvestment in forest ecosystems and the people and communities who depend on them, the community forestry movement has developed a complex institutional infrastructure and a variety of social movement–focused organizations. This infrastructure includes a growing number of nonprofit organizations, professionals, and organizational networks at local, regional, and national levels that provide a diverse array of resources and expertise to the movement. It also includes relationships with philanthropic foundations that have provided significant support for community forestry, particularly those working to change the way society thinks about resources, communities, and the relationship between the two. State and federal agencies are also part of the movement's institutional infrastructure because of their roles facilitating policy and legislative initiatives. Agencies have proven important because of their efforts to build community capacity directly by creating forums for dialogues and through grants to groups for projects. When the grassroots component of the movement, and by this we mean the vast array of nonindustrial forestland owners, forest workers and nontimber forest products harvesters and collectors, and communities enfranchised by virtue of place, are considered, the community forestry movement takes on an exciting and robust appearance. As a national social movement, community forestry has much in common with the sustainable development and communitarian movements and other contemporary domestic social movements that emphasize community-driven policies and programs. It also shares points of similarity and contrast with international community forestry experiences, which offer important lessons for the community forestry movement in the United States.

Diverse Community Forestry Origins and Forms

Although linked by some of the key community forestry themes congress participants identified, community forestry in the United States has emerged in different places, in different ways, and at different times. These diverse community forestry origins and forms reflect the variety of property rights regimes, cultures, social histories, values, economic conditions, and forest visions of those who live and work in the forest. The differences are also the result of the diversity of ecological conditions, including forest fragmentation, the historic and current threat of fire in the forest, and general forest ecosystem health. Although the diversity of community forestry origins, forms, and themes defies attempts at regional delineations, it is possible to identify three broad regionally based categories of community forestry: the Northeast; the Midwest, Appalachia, and the Deep South; and the western United States. These categories emerge from the interaction between the dominant models of forest management and regional patterns of land ownership, demographic and economic pressures, and specific historical and cultural trajectories. Whereas the Northeast and Western United States are conventional regional categories, the Midwest, Appalachia, and the Deep South are rarely lumped together. However, for the purposes of this discussion, at least three factors warrant grouping them. First, there is a pervasive and strong link between agriculture and forest management; forestry often is one component of an integrated working landscape, which includes animal husbandry and crop production. Second, in all three of these areas there is a large and increasing number of small family-scale ownerships. Third, there are many nonindustrial forestland owners, practitioners, and gatherers in these areas who face similar economic pressures resulting from the diseconomies of small scale combined with the fact that many of the communities in the three areas have historically been underserved by state and federal government assistance, extension, and outreach programs.[2]

This threefold categorization facilitates understanding of the diverse initial conditions from which community forestry springs and some of the dominant community forestry forms and themes that have emerged over the last two decades. Although the regional patterns described here are coarse and therefore conceal significant intraregional variation, they nevertheless characterize the dominant social and ecological features to which community forestry has responded. By the same token, these features have significantly affected the regional forms, meaning, and core themes of community forestry.

Northeast

> Community forestry involves the community participating more directly in the management of working landscapes in an ecologically sound way. It reinvigorates historical connections to the land, brings forests and people together, and raises respect for workers and what

> the land provides. It involves a different power distribution. It brings value back to the logger and the wood products industry; and it seeks to change loggers' life expectancy. (Practitioner comments from the region.)

Community forestry in the Northeast has developed in response to a predominantly bifurcated private forestland ownership system. A dual forest management regime has evolved, consisting of large-scale intensive industrial forest management in the northern part of the region (e.g., Maine and New Hampshire) and small-scale, often nonindustrial forest management regimes that dominate in New England. In the industry-dominated part of the region, community forestry appears nascent. Individuals and organizations are concerned primarily with low-impact forestry practices. Others are concerned with labor and forest worker issues, particularly worker safety and compensation, and the need to develop more value-added processing capacities in the region. The negative effects of the North American Free Trade Agreement (NAFTA) are also evident here. Unlike in the West, where cheap Canadian imports had depressed lumber prices before the Bush administration slapped tariffs on them in 2002, whole logs are exported to Canada. Canadian mill workers, who benefit from the social programs of the Canadian government, exert downward pressure on North American labor market wages, and continue to maintain a higher standard of living than comparatively paid U.S. workers.

In New England, community forestry addresses issues of economic viability, forest fragmentation, and interjurisdictional and cross-ownership stewardship regimes within a context of a patchwork of forest ownership patterns. Urban residents' migration to rural areas either full-time or part-time (as second home owners) has increased forest fragmentation and reduced harvests due to new resident opposition to logging. The economic viability of logging operations is threatened as they have an increasingly difficult time securing timber because of forest fragmentation and changing values of owners. With changing values and increasing pressure on remaining forestland, municipal and town forests managed by counties, towns, or cities, particularly those associated with reservoirs, have taken on increasing importance as part of a working landscape.

These general conditions have spawned a diverse array of community forestry activities and forms, many of them recent. In much of the region community forestry focuses on the needs of forestland owners, the increasing threat of forest fragmentation and parcelization, the relationship of forests to both watersheds and community water supplies, and the potential for municipal forests to serve as a model for communal forms of forest ownership and management. Loggers' guilds promote socially and ecologically sustainable logging to respond to the concerns of new landowners, and formally and informally organized networks of nonindustrial forest landowners have developed that promote low-impact forestry and joint forest product processing and marketing activities. County foresters, state forestry organizations, and other forest stewardship organizations are the primary catalysts for community forestry in this region. They promote forest stewardship, landowner education, value-added processing initiatives, and ecologically and economically sustainable forest management. The rich historical legacy of municipal and town forest traditions is enjoying a renaissance as communities realize the value of municipal and town forests in stemming the tide of forestland development, fragmentation, and parcelization. The Beaver Brook Association in the Merrimack River Watershed is one such example. There, on some 2,000 acres of southern New Hampshire and northern Massachusetts land, selective logging of trees to produce income for association activities, recreation, and wildlife habitat is practiced amid a sea of housing developments. To demonstrate forestry practices, the association offers a variety of education programs attended by thousands every year (Lavigne 2003). Municipal forests and their institutional trappings, such as tree wardens and town hall meetings, also provide useful blueprints and contribute to social and human capital that may be extended for the management of community farms and forests (Donahue 1999:xv).

Because of the preponderance of private lands in the region, the federal government's role is less visible than in regions with extensive public forestlands. Nonetheless, the federal government provides significant funding and research support to state, county, and other regional entities. One illustration of this is the Forest Service's Forest Legacy Program. This provides federal funds for incentive-based programs designed to protect important forest areas from fragmentation and parcelization. Using approaches such as conservation easements, multijurisdictional regional stewardship areas, voluntary deed restrictions, covenants, and full-fee purchases, the Legacy Program (whose annual budget increased from $7 million in 1999 to $60 million in 2001) works closely with state and regional partners. These partners identify important forest areas, provide assistance in developing forest stewardship plans, and, against a rising tide of people moving to the more rural hinterlands, help prevent forest fragmentation and maintain working forestland.

The Deep South, Appalachia, and the Midwest

> Community forestry focuses on developing capacities for improved public input into forest planning and the links between that and local, participatory democracy. Community forestry builds alternatives to interest group politics.

> Community forestry cooperatives are comprised of more than landowners; they include practitioners such as foresters, loggers, sawmill operators, and woodworkers, those who "have the feel for it." Land ownership is not the common denominator.
>
> A big problem is the "disconnect" between government and communities of color. A key issue is the need to build trust between minorities and the government. Once agency folks and landowners are brought to the same table, things take off from there almost by themselves.
>
> Community in the South is based on the common ground that we've all been treading on. It is not so place-based as elsewhere, not politically or geographically bounded. It is more comprised of people with shared experiences of oppression, with similar types of problems, and sometimes shared kinship. It is a spiritual community that is very real. (Practitioner comments from these regions.)

The Deep South, Appalachia, and the Midwest share some similar forest conditions but are obviously quite different historically, socially, and culturally. The boom-and-bust natural resource extraction cycles that figure so prominently in the West for the most part have been absent from these areas for many decades. In the 1980s and 1990s, these areas generally did not experience the sudden economic downturns and paralyzing conflict and gridlock that the West had. The majority of forestland in these three areas is privately owned. All three areas are characterized by large numbers of nonindustrial forestland owners, many of whom practice diversified land management and integrate forest management with agricultural activities. In Appalachia and the South, large-scale corporate agriculture and forestry operations challenge nonindustrial private landowner economic viability. The struggles of many resource-limited small landowners in the southern areas to keep afloat are compounded by the fact that historically they have also been underserved by state and federal extension agencies and by the extension programs and research agendas of the land grant universities. Institutionalized racism has conditioned the flow of government support (subsidies, loans, expertise, extension) away from African American landowners and forest practitioners. Public forests in these regions, which are primarily national forests, are characterized by much of the same policy conflict and gridlock prevalent in the western United States.

The Midwest and the South also have large-scale industrial forest management operations. As with industrial forestland ownerships in the Northeast (and elsewhere in North America), community forestry opportunities in these ownerships revolve around worker issues, recognition of customary rights, and nontimber forest products collection. The large and increasingly consolidated industrial forest ownerships in the Deep South, part of the new woodbasket of the country, tend to be disinclined to recognize the historical customary rights of local communities to those forestlands, such as bear hunting in bottomland forests. Furthermore, all too often industrial operations employ migratory work crews under conditions that may not or may only barely meet minimum required working conditions and pay.

In the South, Appalachia, and the Midwest, community forestry springs from the concerns of smaller nonindustrial forestland owners and agriculturalists, forest practitioners, and non–forestland-owning communities who use forests for collecting nontimber forest products. These communities include a disproportionately large number of underserved and resource-limited landowners. At least in some rural parts of the Deep South, Appalachia, and the Midwest, entrenched poverty, lack of economic opportunity, and decreasing returns to labor have depleted community stocks of financial, human, cultural, and social capital and encouraged migration, especially for youths and young adults. Nonprofit organizations in these regions, with support from foundations and state forestry programs, sponsor forest, farm, and business management workshops, help minority and limited-resource landowners network to pursue common interests, and help improve landowner access to the government services and markets that have historically been unavailable to many small nonindustrial forestland owners, especially minority and limited-resource owners.

Community forestry in these areas is strongly rooted in traditions of social organizing, outreach, and the mobilization of collective interests. In the Midwest, community forestry is rooted in the region's strong tradition of cooperatives, which serve as a powerful vehicle for mobilizing and realizing the collective interests of communities of forest owners, practitioners, and users. The Western Upper Peninsula Forest Improvement District in Michigan, encompassing 200,000 acres and hundreds of landowners, is one such example (Mitsos 2003). Many of these cooperatives emerged from the grassroots. They were formed by forestland owners as well as forest practitioners, including loggers, sawmill operators, and woodworkers. For many landowners, especially those who are farmers and similar to their counterparts in the South, forestry had been a hitherto underused component of their overall land management strategy. Nonprofit organizations such as the Federation of Southern Cooperatives provide technical forest planning and management assistance through seminars and training programs, facilitate the development of landowner, practitioner, and gatherer cooperatives, support forest certification for small landowners, offer financial planning and marketing assistance, and serve as a bridge between these communities and state and federal agencies concerned with forest management and agriculture issues.

The West

> Community forestry developed because people are tired of fighting in the West. Community forestry involves people interacting with forests; it entails communication across divides. Community forestry needs a "thicker" definition of community, one that attends to issues of class, race, ethnicity, gender, and history. It must have authentic people's participation. Community forestry involves securing access to training, education, jobs, and contracts. It needs to more broadly define community to be inclusive of all groups. (Practitioner comments from the region.)

There is an old African proverb that states, "When the elephants fight, the ants will get stamped on." One of the primary conditions that gave rise to community forestry in the West are the fierce interest group battles that had no regard for locals and left rural people feeling stamped on. The winner-take-all battles between national environmental groups and the timber industry over access to federal timber, and the associated policy and management gridlock, left rural communities with few options. Some residents blamed environmentalists for the repeated court appeals over timber sales and forest plans that delayed or stopped work altogether in the forest. Environmental group successes contributed to the dramatic reductions of timber harvesting on federal land, an economic lifeblood for many Western rural communities in a region where vast acres of forestlands are public, managed for the most part by the Forest Service and the Bureau of Land Management. Other residents blamed the timber industry for its long-standing reliance on big trees and company practices that left places with degraded forestland, abandoned mills, and impoverished communities. While the national environmental movement's influence in the West grew over the last two decades of the twentieth century, the wood products industry's access to federal forests declined. During this same period the wood products industry in the West was confronted with an increasingly competitive global industry, cheap imports, older, inefficient mills that relied on old-growth timber, overcapacity of existing mills, and increasing industry concentration. Coupled with the reduced access to federal forestland, these factors conspired to push wood products industry activities out of the rural areas that relied on them most and out of forests that needed attention.

While the elephants battled, the people who lived in the rural communities often battled among themselves and struggled economically, socially, and spiritually. The struggle in the 1990s went beyond access to timber for jobs; it involved maintaining vital communities and healthy forests. For many across the West, access to the forests that surrounded communities was needed to reduce the extreme fire risk caused by fuels buildup and drought. The dominant patterns of timber harvesting and associated fire exclusion policies in the West created forest stands characterized by suppressed small-diameter high stem densities, excellent targets for insect infestations and stand-replacing fires. For Hispanic groups in New Mexico, the issue was even more basic: It was about securing access to land (that was formerly their own) to obtain firewood to heat their homes through the winter. For communities across the region, it was also about healing the deep rifts associated with the interest group battles. As one member of a rural northern New Mexico community described, "When we realized our children were growing up hating the kids of the other side, community members realized they had to find a way out of gridlock," thus igniting another locally based set of community forestry efforts and processes.

Community forestry in the western states developed from hundreds of "spontaneous ignitions" across the region in which residents of rural resource-dependent communities and forest workers, sometimes with the help of facilitative local government, universities, nonprofit organizations, agency personnel, and political leaders, sought to creatively and constructively transcend the gridlock, acrimony, loss of jobs, and reduced access to forest resources. The Quincy Library Group is perhaps one of the most well known, as much for its success in bringing contentious local parties together, developing a joint agenda, and having it turned into legislation as for its vitriolic exchange with national environmental groups that led to the failure of the legislation to achieve its desired ends.

Focused predominantly on public land and on forest planning and management processes, community forestry in this region involves grassroots-based, multiparty stakeholders. By *grassroots, multiparty* stakeholders we mean diverse interests—including locals who identify with if not represent national interests—coming together in common cause over place-based issues. These processes often focus on forest and watershed restoration activities, fuels reduction programs, thinning and utilization of suppressed small-diameter trees, the development of markets for historically underutilized species (such as tan oak and madrone), the management and harvesting of nontimber forest products, and other value-added processing activities. The focus of many of these efforts is the restoration of a mutually beneficial relationship between the people who depend on the forest and the forest ecosystem itself. Recognizing that gridlock is detrimental to almost all forest stakeholders, these communities have begun the difficult process of hashing out their differences in an attempt to identify a middle ground agenda that they can collectively support and pursue. Local environmentalists, loggers, forest workers, and political leaders recognize and acknowledge that in their present condition many if not most forests are increasingly vulnerable to stand-replacing fires and insect attacks, forms of disturbance in scale and intensity that appear to lie far outside the historic range of variation.

The dominance of public forestland ownership in the West also means that the federal government, especially the Forest Service, is a central and dominant player in community forestry. Thus issues of organizational flexibility, decentralized decision-making authority, and shifts from centralized hierarchical to more network-like organizational structures are crucial to community forestry debates in this area. Similarly, the relationship between local place-based communities, mobile forest workers and gatherers, and public forest management agencies, particularly the Forest Service, is central to community forestry in the region. One primary element of this relationship concerns the manner in which contracting for work on the forest is structured and who benefits from the contracts awarded, a focus of the work of the Alliance of Forest Workers and Harvesters based in Eugene, Oregon. The history of acrimony between national-level interest groups (environmental and industry alike), public agencies, and local communities has resulted in a lack of trust, and all too often nonlocal workers and nontimber forest product gatherers haven't even been part of the conversation. Despite the fact that Canadian tree planters have developed a strong union, are paid well, and work in what is considered a respected profession, they exert no upward wage pressure on their U.S. counterparts. U.S. tree planters are mostly mobile Latino forest workers and are poorly paid. Discussion of these workforce issues has only recently begun in the West, led by the Oregon-based Jefferson Center, but in neither region is there a developed community forestry movement to address these issues. A challenge of community forestry in the West is to rebuild trust between these groups and to rebuild work and occupation as a basis for forest enfranchisement for both mobile and place-based workers. Community forestry in these areas is the attempt to restore both the forest ecosystem and the livelihoods of people who have depended on them and who have the knowledge to steward them.

Common Unifying Themes

Despite the diverse origins, forms, and themes of community forestry around the country, there are several shared important core features. These shared themes, in conjunction with the movement's infrastructure—various regional and national networks that link grassroots community forestry practitioners, supporters, researchers, and public and private institutions—provide the coherence, stability, and national-level presence necessary to call community forestry a social movement.

One of the most common unifying features of community forestry is the attempt by people to reorder relations among themselves and between themselves and the forests on which they depend in a manner that simultaneously promotes or improves the forest condition and enhances community well-being. In this context, community well-being means much more than maintaining or increasing forest sector employment levels, as earlier Forest Service community stability programs sought to do. Rooted in grassroots and participatory democracy, community forestry is about encouraging bottom-up forms of development and forest management that are conceived, developed, implemented, and monitored by communities themselves, often in partnership with supporting public and private institutions. Such bottom-up approaches entail a reconfiguration of relationships between rural residents and workers and the wider political and economic systems with which they interact. This reconfiguration seeks to empower the communities and workers that historically have been disenfranchised from the forest ecosystems (both as natural and financial capital) on which they depend. Thus a common element of community forestry is the development of community capacity—the financial, physical, human, cultural, and social capital—necessary to develop and guide community-involved forest and ecosystem management efforts that maintain or restore desired forest conditions and promote community and worker well-being.

Closely related to reordering relations is collaboration, another core community forestry theme. Community forestry involves reordering social relations between people in a manner that promotes more collaborative forms of interaction. Community forestry collaboration differs from more general collaboration because of its focus on place and on people who are involved with that place. It is not simply about stakeholders discussing policy distantly connected to the land and the people. As a result, collaboration can take diverse forms. It includes cooperation between landowners and forest practitioners through collaborative groups created to improve the quality of forest management and social and economic outcomes. It involves increased collaboration between minority and limited-resource forestland owners and the government extension and other agencies that have historically underserved them. It includes the explosive increase in collaborative multiparty stakeholder planning processes, most common in areas with significant public land. And it involves processes that typically are open and expansive rather than exclusive and restrictive, as practitioners recognize that collaborative problem solving is most successful when it includes multiple stakeholders and groups at multiple scales, and ecological and social coherence takes precedence over administrative or politically expedient boundaries. For community forestry to be successful in the long run, it must achieve effective and meaningful collaboration among all the diverse groups of people with interests and a stake in the forest, for example: between place-based groups and communities of interest, between place-based groups and mobile workers, among adjacent nonindustrial forestland owners, between non–forestland owning communities and forestland owners, among different groups of nontimber forest product collectors, and between them and forestland owners and agency managers. In many rural communities these types of collaborative efforts also involve increased coordination between forest management and broader community goals related to local economic development and planning.

Investment in natural capital and community economic and social health is another core community forestry theme. Investment concerns stem from long-dominant forest extraction practices that reduce long-term productive capacity and diminish the capability of the forest ecosystem to support diverse uses and provide diverse benefits. A focus on investment involves reversing natural capital drawdowns that simultaneously impoverish forest ecosystems and local communities and workers. Creating jobs, supporting small businesses, and improving the viability of forest landowners have been the foci of many community forestry practitioners. Other examples of investment include developing markets for the byproducts of forest restoration and value-added processing (often through cooperatives or cooperative-like business and marketing ventures), reorienting timber harvesting from exclusively a commodity extraction approach to one in which ecosystem restoration is an important land management objective, developing stewardship contracting mechanisms on national forests, and developing the capacity for the sustainable in situ and ex situ production of nontimber forest products. Establishing a market for carbon sequestration or reinvesting the benefits from the downstream sale of water in the forest watershed are two of the newer and more powerful mechanisms for reinvesting in the system. These issues are discussed further in Chapter 9.

Investment begins with identification and assessment of benefits (monetary and nonmonetary) and with the development of institutional mechanisms and action to ensure that the benefits of natural capital are recognized and paid for. Collection alone is insufficient; returns must be reinvested to maintain or enhance the biophysical or dependent social systems. This may involve redistributing benefits (monetary and nonmonetary) in a manner that provides more benefits to forest owners, adjacent communities, and people who work in the forest. Utilization of the natural capital of the forest must be replenished. Because of long-term natural capital drawdown and disinvestment in communities, reinvestment involves rebuilding the community stocks of social, cultural, human, physical, and financial capital. Reinvestment is a bottom-up process that begins with a focus on both natural capital and community capital. Use of either requires reinvestment. However, community forestry practitioners recognize that the reinvestment issue, though it must be responsive at the community scale, usually can be advanced only through work at political and institutional levels beyond the community.

Implicit in several of the themes discussed in this chapter is the notion that community forestry entails a variety of institutional changes at multiple levels. Within public land management agencies community forestry on public lands requires new institutional relationships. This includes changes in the budget process that counter the historical link between budget allocations and commodity outputs; changes in the organization and packaging of work on public lands; changes in planning processes to make them more participatory and democratic; changes in the way science is conducted and how knowledge is generated, exchanged, and used to make decisions; and changes in attitudes and programs for fire and fuels management. In general these changes entail decision-making procedures and forms of organizational authority that are more decentralized and less hierarchical. On private lands community forestry entails institutional changes with respect to how natural capital is valued and taxed and the restructuring of access to technical and financial assistance for nonindustrial forestland owners. It also includes developing multilandowner organizations that seek to coordinate land management planning and management across ownerships and develop value-added forest product processing opportunities. Community forestry involves extension and outreach agencies diversifying their vision of the forest (and concomitantly of extension needs) to correlate more closely with that of the broader community of forest users and owners with whom they must now interact more closely than before.

Community forestry also challenges traditional Progressive Era science and the dominance of silvicultural prescriptions for forest management based on the Germanic model of scientific forestry. Community forestry practitioners argue the relevance of local knowledge for forest management and embrace a pluralistic attitude that validates the importance of both local knowledge and scientific knowledge and stresses the importance of integrating them. This is especially relevant in the context of forest management goals that include forest restoration and the management of nontimber and nontraditional forest products. In many respects community forestry challenges the mainstream practice of science. One potent example involves mushroom harvesters picking and monitoring matsutake mushrooms in southeast Oregon. Through a nongovernment organization–inspired and foundation-supported monitoring program, harvesters learned that some of the best matsutake-producing areas on national forestland were slated for timber harvest. Not only had the matsutake mushroom production not been considered by the Forest Service, despite the fact that its value exceeded that of the timber, but the National Environmental Policy Act analysis, which required documentation of the socioeconomic effects of the timber harvest, ignored the mushroom-destroying effects of timber harvest and the millions of dollars of lost income by mostly low-income harvesters.

Community forestry practitioners and supporters argue for the principles of civic science in which communities themselves participate in scientific inquiry, from the identification of research questions to the research design and monitoring of results. Community forestry practitioners argue that information and multiple forms of knowledge should flow from scientists and the agencies to the public and back, quite different from the one-way flow that has characterized expert–public interaction. In this respect community forestry represents a significant challenge to the Progressive Era faith in science and the "gospel of efficiency" (Hays 1959) as the best way to identify and achieve the public interest.

Perhaps the last emerging theme is the extent to which community forestry embraces the multicultural ethnic and racial diversity of forest owners, users, and workers. If community forestry is to achieve its claim to be a movement for social change that empowers people who have hitherto been disenfranchised from the forests on which they depend and excluded from dignified and full participation in the forestry sector, then community forestry practitioners, supporters, and advocates must seriously consider the historical legacy of community and worker disenfranchisement from forests and plan how to overcome that legacy. For many, community forestry has a social activist core. For all, its long-term sustainability depends on its ability to meet the challenge of incorporating the needs and visions of the full spectrum of forest stakeholders. Developing ways to manage forests that are inherently multicultural and inclusive of a wide variety of values, interests, and stakes in the forest is a challenging process that entails transcending old divisions and creating decision-making and planning contexts characterized by tolerance for diversity and trust. Although challenges remain with regard to these issues, the enfranchisement and empowerment that have already occurred through community forestry are a powerful beginning.

A Unified Vision of Community Forestry

A number of the themes described in this chapter have been drawn together in a unique set of community forestry principles developed by the Lead Partnership Group, a group of some 20 community-based forestry and watershed groups from southern Oregon and northern California.[3] This group was launched by one of the authors and Forest Community Research after the Clinton administration's release of the Northwest Forest (Option 9) Plan. Apart from the dramatic reductions in timber harvests for which the plan is perhaps best known, the Northwest Forest Plan established experimental management areas in the national forests called "adaptive management areas" in which experiments in community–forestry interactions and land management with more community involvement and benefit could be advanced. The Lead Partnership Group was formed to offer a community voice to the agencies and institutions responsible for implementing adaptive management areas and the practice of adaptive management itself, another hallmark of the Northwest Forest Plan. Adaptive management calls for more conscious and immediate learning from ongoing management that informs and guides subsequent management action. Mistakes are welcomed as learning opportunities that iteratively inform and lead to improved management. What makes adaptive management different is that it requires those managing to be more conscious about learning as they manage and to quickly incorporate lessons learned into management action. Some have described the adaptive management process as "managing to learn and learning to manage." In what may seem like an obvious approach to resource management, developing the adaptive management capacity of agencies such as the Forest Service, which for years focused on the volume of timber produced rather than the condition of the land and for the last 20 years had been focused on intensive planning and environmental analysis before doing any on-the-ground management, is no small task. The Lead Partnership Group provided much input to federal agencies responsible for implementing the Northwest Forest Plan in attempts to help the agencies understand how to manage resources in more socially and economically responsive ways. It ultimately proved to be frustrating because of the entrenched institutional barriers to adaptive management resulting from the long-standing commitment to timber production and a Progressive Era–inspired reliance on dispassionate and detached scientists and managers who themselves lacked understanding about how to effectively include public perspectives, much less actively engage with community constituencies.

Despite their frustrations with the Northwest Forest Plan and adaptive management, the Lead Partnership Group continued to meet because of what they learned from one another and the ideas they jointly incubated and because collectively they could more effectively tackle institutional issues that affected them all. The group spent a year developing papers on core community forestry themes such as monitoring, forest health, stewardship, and reinvestment. Five years into its meetings, the Lead Partnership Group agreed to develop principles of community forestry, and for a year and a half, they worked on them. They started with ideas first generated in the papers they produced on core community forestry themes and expanded them based on what they learned since the papers were produced. In the process of developing and refining the themes, they disagreed and argued with one another, but the groups persevered and ultimately produced a powerful set of principles (Box 4.1) that reflect many current unifying themes of community forestry with applicability to areas far beyond the Pacific West.

Box 4.1 Lead Partnership Group Principles of Community-Based Forestry

Preamble

This document contains principles of community-based forestry developed by groups that make up the Lead Partnership Group. The principles that follow are based on our experience and have served as guidance to many of our groups. They may also serve as guidance for groups addressing other natural resource issues. They are indeed principles (motivating forces, rules of conduct, or essential elements) but should not be considered a litmus test to determine whether a group is a community-based group. At any given time, groups may find that their actions demonstrate all, some, or few of these principles. Because all places differ ecologically and socially, of greater importance should be the continued effort to find common ground and to work for supportable, sustainable, and locally sensitive solutions.

What Is a Defining Quality of a Community-Based Group?

Community-based groups, as discussed in this document, are those that attempt to find realistic long-term, sustainable solutions to resource conflicts and ecosystem management through stewardship and by implementing practices that combine local knowledge with the best science. These solutions are generally understood and acceptable to local residents.

Process

1. Resource management must be guided by a commitment to environmental health and social well-being.
2. Consensus-based processes and decision making are central to community-based groups; where consensus does not exist, other civil democratic processes must be used.
3. Multiple and public stakeholders—those who identify themselves and their interests as being linked with forest ecosystems—have the right and responsibility to be involved in forest management.
4. Community-based groups have the continuing responsibility to encourage a broad diversity of interests within the group.
5. All communities, whether place-based or interest-based, must recognize that all interests (including local communities) have the right to access and to use decision-, rule-, law-, and policy-making processes.
6. Communities of interest have a legitimate interest with place-based processes: Their participation in community-based

continued

Box 4.1 continued

processes has the potential to improve social, economic, and environmental health in addition to fostering agency, industry, and community accountability.

7. Management and regulatory agencies should incorporate place-based and interest-based knowledge, skills, and perspectives into their planning, implementation, monitoring, and evaluation.
8. Workers' issues are an integral part of community forestry. Forest and watershed workers have the right to living-wage compensation; decent working conditions; and worker-accountable representation in affairs that affect their health, safety, and terms of employment. They also have a responsibility to participate in the protection and enhancement of ecosystem integrity.
9. The social, educational, economic, and environmental benefits derived from community-based forestry should be produced in a manner designed to increase the capacity of local communities of place so as to maintain those benefits for local, regional, and national communities.

Policy and Institutions

10. A. Community-based groups will uphold environmental laws and regulations and work to improve environmental and social justice.

    B. Future environmental laws and regulations should be established with the active involvement of multiple stakeholders—urban and rural, at the national and local levels—while considering international communities.
11. A. Public lands should remain public and be managed for long-term sustainability.

    B. Government agencies whose mission it is to manage these lands should improve their performance and potential as national stewards.
12. Ecosystem management must be responsive to both private and public rights and responsibilities.
13. When communities of place and communities of interest agree, federal agencies have an obligation to facilitate implementation. When there isn't agreement, agencies have a responsibility to facilitate democratic and transparent processes to make decisions.

Monitoring

14. A. The choice of monitoring process affects social well-being and environmental health. All-party, multiparty, or third-party monitoring that is open to all should be adopted as a managerial standard.

B. Adaptive management should be implemented as a managerial standard. Monitoring that is open and accessible to all is essential to improve learning and future management.

C. The effects of land management, active or passive, must be monitored and evaluated at multiple temporal and spatial scales.

Stewardship

15. Forest-based industry that is both responsive and accountable to the needs of forest ecosystems and forest communities is necessary and valuable.
16. A. Local knowledge (experience, information, expertise) and perspectives together with sound science can improve management.

    B. Local residents of forest communities are intrinsic to effective forest stewardship and must have the opportunity to be involved in forest planning, evaluation, monitoring, and restoration.
17. It is in the regional and national interest to support local capacity for stewardship.

Reinvestment

18. A. Healthy ecosystems benefit society. Reinvesting in maintenance and restoration of ecosystems is the responsibility of all beneficiaries.

    B. Distant users through government and market mechanisms must also be responsible for reinvestment in maintenance, restoration, and remediation of ecosystems that produce clean water and clean air from which they draw shared social and environmental benefits.

    C. Exclusive reliance on user fees to maintain ecosystem health is insufficient and inequitable; it changes the fundamental relationship between the public and public lands and should be avoided.

The Lead Partnership Group's principles of community-based forestry have since been shared with groups around the country and have been broadened to principles of community-based resource management. They continue to inform the work of individual groups and have helped others understand how these leading western groups view community forestry. Unable to practice community forestry in adaptive management areas, groups proposed the idea of pilot projects, an idea now enshrined in the national stewardship pilot projects and implemented in 28 pilots in national forests across the country. Recognizing the critical importance of monitoring, long underfunded and for the most part ignored, seven groups individually developed projects to advance all-party monitoring pilot projects. The groups launched these projects to engage regional and national interest groups in their own work and to advance understanding of how to implement it. Inclusion of environmental groups is important because regional and national environmentalists have grown increasingly concerned that they do not have the time and cannot afford to participate in all emerging community forestry projects.[4] Environmental groups are not the only groups invited to these processes. Groups invited scientists, public agency representatives, and other stakeholders to establish assurances that local resource management projects would be responsive to stakeholder concerns. Approaching its 10-year meeting anniversary in 2003, the Lead Partnership Group has grown into one of the most advanced community- and natural resource–focused ongoing conversations in the country. Yet it is not alone. Since the group first started, there has been a dramatic growth in the number and kinds of groups focused on community forestry, including groups such as the Four Corners Partnership and the Healthy Forests, Healthy Communities networks focused on regional issues, trying to change social and institutional structures to more effectively and equitably improve forest resource valuation and allocation.

The Movement's National Infrastructure

The themes, visions, goals, and challenges shared by community forestry practitioners and supporters across the country provide the basis for an emerging national-level movement. The movement is building from local, on-the-ground work and local successes. Increasingly, it is reaching up and out through regional dialogues and networking and is becoming institutionalized at the national level through events, such as the Seventh American Forest Congress and other conferences and meetings, and a few national-level organizations and networks. Community forestry's national-level presence helps to identify the commonalities that underlie regional variations in ecology, economy, and society. This facilitates the forceful articulation of the common constraints, challenges, and opportunities within community forestry, a necessary precursor to developing a common vision for social and environmental change. Community forestry's national presence also advances the movement by increasing its visibility in state and national policy-making arenas and attracting foundation and philanthropic support.

The National Network of Forest Practitioners, with a member base of 450 people, is one of the most prominent national-level community forestry organizations. The organization facilitates exchanges between community forestry practitioners across the country, advocates national policies that favor community forestry and its practitioners, and sponsors forums in which practitioners can engage with policy makers at state and federal levels. The National Network of Forestry Practitioners' efforts to promote collaborative and participatory community forestry research culminated in the creation of the National Community Forestry Center in June 2000, made possible by a grant from the U.S. Department of Agriculture's Fund for Rural America. The National Community Forestry Center is the only nongovernment organization funded; the other four Fund for Rural America Centers are housed at land grant institutions. The broad aim of the National Community Forestry Center, which is a decentralized network of regional centers in the Southwest, Southeast, Pacific West, and the Northeast, is to "improve the access of people . . . to research and researchers, to build their research capacity, and to involve them in the research process." These four regional centers represent a few of the many regional organizations that have emerged over the last decade or so whose mission includes advancing the community forestry agenda at the national level.

Perhaps the other primary national community forestry–focused organization is the Seventh American Forest Congress Communities Committee. Less a group than a prominent network of organizations and individuals across the country, the Communities Committee of the Seventh American Forest Congress was organized with the express purpose of promoting community involvement in forest management and highlighting the linkages between responsible forestry practices and community well-being. The committee works closely with American Forests, the Pinchot Institute, and the National Network of Forest Practitioners to advance national policy. For example, in the final months of 2001, much effort was devoted to advancing community forestry issues in the Farm Bill.

All of these groups work closely together to organize and support community practitioner visits to Washington, D.C. Known as the Week in Washington, these visits consist of meetings with key congressional staff, Forest Service and other federal officials, and national interest groups. American Forests staff members, who are based in Washington, D.C., and play central roles in planning the visit, strive to time the event to coincide with congressional hearings on issues related to community forestry. As a result, participants have had the opportunity to provide testimony to congressional committees. The overall experience, often practitioners' first visit to Washington, D.C., is a crash course in national policy making and advocacy training. The visits provide important opportunities to demystify Washington politics for practitioners, who are often based in small rural communities; for many, this is their first opportunity to learn first-hand about the national institutions and processes associated with democratic pluralism. Equally importantly, the

Week in Washington visits have helped raise congressional awareness of key community forestry issues, and practitioners often return to Washington to testify on forestry or forestry-related bills or to work with the agencies on community forestry issues. For example, recent work on the National Fire Plan has drawn a number of community forestry practitioners to the Capitol, some numerous times.

Many civil servants also work to advance community forestry. Examples include extension foresters who work with nonindustrial forest landowners, rural development specialists within the Forest Service (especially the Division of State and Private Forestry), and other line officers in public lands agencies who are committed to working collaboratively with rural stakeholders. Though limited in number, researchers and practice-oriented personnel in colleges and universities also play important roles in facilitating community forestry initiatives, helping to identify relevant research questions, and designing and conducting research and monitoring of community forestry projects.

Finally, the community forestry movement in the United States owes much to two different but somewhat interrelated sources of support. The first is the committed financial support of many philanthropic organizations interested in promoting innovative grassroots-based efforts to develop integrated solutions to environmental and social problems. The Ford Foundation has committed substantial support for a selected pool of community forestry organizations through its national community forestry pilot demonstration program, and it supports graduate research on a variety of issues through its community forestry research fellowship program. Other foundations that have provided generous support to the community forestry or important dimensions of the movement include the Surdna Foundation, the William and Flora Hewlett Foundation, and the James Irvine Foundation.

The intellectual and inspirational support that international community forestry provides to the domestic community forestry movement is an important complement to the financial support foundations provide. Indeed, these forms of support are in some cases interrelated because it is not unusual for foundation program officers working to advance community forestry in the United States to have previously spent long periods of time engaged with community forestry in other countries. Similarly, some community forestry practitioners have had the opportunity to meet and share experiences and insights with community forestry practitioners and supporters in other countries, and representatives from other countries have spent time in the United States meeting with members of the domestic community forestry movement. Although a comparison of the similarities and differences between community forestry in the United States and other countries is well beyond the scope of this work, the parallels are abundant. This is partly because the dominant model of forestry described in Chapter 3 prevailed in other countries in addition to the United States. Thus, the various community forestry movements that have arisen around the world in the last 20 to 30 years, in many respects, are responding to a similar set of constraints, barriers, and opportunities.[5]

Conclusions

Does community forestry constitute a movement? Based on the definitions of social movement with which this chapter began, it seems clear that community forestry is focused on social change and has the integrity of vision, internal coherence, institutional infrastructure, and access to resources to be called a movement. Community forestry practitioners, supporters, students, and advocates seek the changes in social structure and allocation of value described in this chapter. Moreover, there exists a robust set of social movement organizations related to community forestry at the grassroots, regional, state, and national level. Although they are necessarily diverse, their diversity suggests the wide array of interests to which community forestry appeals and its underlying unifying themes. When facilitative public agencies and civil servants, supportive philanthropic organizations, and the rich array of international sister community forestry movements are considered, it becomes clear that there exists a robust, innovative, and exciting domestic community forestry movement that is focused on achieving the interdependent goals of stewarding forest ecosystems and improving community well-being. Chapter 5 explores the central objectives and goals of this movement and presents a way of analyzing both the problems it seeks to ameliorate and how it may do so.

CHAPTER 5

Environment, Economy, and Equity

> At the very time when advanced industrial societies such as the United States have developed a concern for and an appreciation of the natural environment, they are allowing the social environments that make human development possible to deteriorate.
>
> —*Wolfe (1995:163)*

This chapter marks a substantive shift in the nature and thrust of this book. The prior chapters discuss the historical antecedents of community forestry, review dominant historical forest management trends and the resulting conditions that gave rise to community forestry, and provide an overview of the diversity and unity of themes across the country that define the emerging community-based forestry movement. In this and the next four chapters, we turn our attention to an analysis and assessment of the current community forestry movement itself and identify its future challenges. We suggest that the triad of environment, economy, and equity captures the overarching objective of community forestry: to develop new relations between people and the forests on which their well-being depends and that maintain or enhance ecosystem processes, generate sustainable streams of revenue for investing in ecosystems and communities, and promote democratic values of civic participation and self-determination. We use the capital assets framework to frame the core system in which community-based forestry practitioners work. This framework discusses the human community and workers and the forest resource in terms of capital assets. The relationship between the community, workers, and natural capital—the forest—is mediated by institutions that determine both the opportunity and the likelihood that new relationships between people and resources develop and if and how essential investments in forests and human communities are made.

This chapter also marks a shift in the voice used in this book and the material on which we base our comments. Chapters 2 and 3 are based primarily on analysis and review of primary and secondary literature; Chapter 4 relies on a combination of primary and secondary literature along with information from the workshops and interviews with community forestry leaders from around the country. By contrast, in this and subsequent chapters, our perspective shifts from that of observer toward a more activist engagement with the material; our motivation expands beyond seeking to understand to advancing and strengthening the equity and social justice meanings of community forestry and our understanding of the fundamental processes associated with it.

Our analysis and assessment of community forestry include both what community forestry is and what it could, and perhaps should, become. Much of this is explored in our discussions of integration, the environment–economy–equity triad, and the capital assets framework, all in the context of institutional barriers, incapacities, and opportunities. The underlying emphasis on the importance of embracing more fully the potential of community forestry to address equity and social justice issues stems from our perception that the formal community forestry movement has made great progress in a short period of time with regard to environment and economy but that similar progress has not been made with respect to equity. Some groups and communities whose well-being depends on the forest and who fall within the purview of community forestry have not yet engaged with community forestry either because they lack the capacity to do so or because they do not feel that the movement is responsive to their needs. The view that equity is the stepdaughter of the community forestry movement and has yet to receive the attention it deserves evolved during this study. Given the uneven playing field that exists, the years of disinvestment that have impoverished people, communities, and the landscape and the resulting heightened competition for scarce resources, and the multicultural composition of the communities involved, it is little wonder that community-based forestry is not yet as inclusive as we hope it becomes.

Place-Based Integration of Environment, Economy, and Equity

Community forestry, like other examples of the new localism such as civic environmentalism and the sustainable communities movement, highlights the importance of place in the development of a group's collective identity. It asserts that being grounded in a place enables people to build community and strengthen civic institutions that promote citizen participation. It is also fundamentally about validating and revitalizing the relationship between people and the environment that surrounds and sustains them. This revitalization occurs through myriad ways. One is through the development of a stewardship ethic that springs from the recognition that community health is inextricably linked with the health of forests. The concept of "ecological poverty" (Agarwal and Narain cited in Oliver et al. 2000) is developed from a recognition of the strength of this relationship. The concept holds that human impoverishment is the direct result of landscape degradation. Improving conditions for the poor entails restoring the land, which takes capital investment beyond that of the labor input of the poor. Additionally, when natural capital is seriously depleted, improvement cannot be based on market mechanisms; instead, collective or coordinated actions are needed. Recognition of this relationship yields efforts to steward and preserve the ecological integrity and productivity of the working landscapes on which communities and workers depend.

Local knowledge and its application in managing working landscapes is a core component of community forestry initiatives. This involves creating mechanisms that validate, apply, and strengthen local knowledge concerning ecosystems and how they may be best managed to ensure their own sustainability and that of the communities that interact with them. Place is the field in which people and knowledge are rooted as landscapes encode and embed social histories. Making visible and understanding the diverse groups and communities that have gone before, whose actions have in large part created the current landscape, leads to an understanding of the short duration of current communities' "ownership" of resources and emphasizes that what we claim as ours is in our possession for only a short period of time and therefore must be stewarded for future generations. The Menomonee Tribe in Wisconsin and their practice of forestry for the seventh generation is one example of this. Another implication of revealing the diverse cultural histories encoded in most landscapes is that doing so heightens awareness of the current diversity of communities that interact with natural resources and of the validity of their claims to those resources. Gary Nabhan (1997:294), discussing the "new" conservation movement that resulted from a gathering of Native Americans, African Americans, and Hispanics to discuss an international agenda for environmental justice, said, "It was clear that a major shift had taken place in the conservation movement. It is a shift toward inclusiveness and away from the heroic actions of a select few. It asks us to listen to the many voices associated with the land—to learn from them, to celebrate them . . . and to safeguard their legacy." This perspective reduces the tendency to think of forest ownership in absolute terms and encourages the notion of ownership as a bundle of rights to which there may be diverse and legitimate rights holders and appurtenant to which are long-term stewardship responsibilities.

Community forestry is at heart a place-based, integrative enterprise that may be thought of as a three-legged stool. The three legs—environment, economy, and equity—are equally important components. Unless all three legs are sturdy, the stability of the stool itself is threatened. This integrative approach has grown out of a battle of interests that dichotomized the problems of the forest. Responding to agencies and an industry whose objective was "to get the cut out" with little regard for the long-term health and sustainability of the forest, environmentalists challenged first clearcutting and then later other forms of timber harvesting that degraded the forest. In a movement that itself came of age in the 1960s,[1] environmentalists importantly changed the thinking and discourse about natural resources and changed how people think about forests, including the multiple benefits they provide. However, environmental group concerns for forest ecosystem sustainability were not matched by commensurate concerns for the people and communities affected by declines in the forest resource industries or environmental policy.

When not blatantly supportive of one interest over another, forestry professionals and managers responded to the industry–environmentalist battles by calling for a balance of competing interests, whether it involved balancing the number of acres clearcut relative to how much slope and soil instability was tolerable or harvesting more timber at the expense of ensuring high-quality habitat or species protection. "Balancing" interests in resource management, a managerial formulation enshrined in the Multiple-Use Sustained Yield Act of the 1960s, led to zoning in federal forests, a practice in which areas are dedicated to a primary use. On private land, "balancing" required owners to develop harvest plans more responsive to increasing environmental restrictions, with the degree of "balance" decided on a state-by-state basis. The battles associated with forest management across the country have broadened, but the dominant interests that are balanced still involve the natural environment with the wood product industry and, more recently, the natural environment with development. The result, on federal land, for example, is a zoning that reflects more the existing interest group divisions and groups' respective power than ecological imperatives and human and community well-being. Similarly, for private landowners, the "responsiveness" required reflected less the condition of the forests and more the politics and power of environmental and industry lobbies at a state level.

One of the worst expressions of balancing is captured in the phrase "owls versus jobs." Advanced first in the Pacific Northwest, the phrase suggests that more of one means less of the other. This phrase grossly oversimplifies the challenge of managing forests to support spotted owls (and the forest complexity for which owls were to serve as a proxy) and ignores the myriad factors that have led to and will continue to contribute to the decline of industry jobs. The phrase makes for a great sound bite, but in its glibness it dichotomizes and divides, sharpening the interest group battle lines and narrowing the thinking and exploration of options that are so vitally needed to address inevitable "imbalances" that result from resource conflicts, new knowledge, and changing values. Discussing the limitation of interest group politics, Bellah (1995:53) states, "The politics of interest provides no framework for the discussion of issues other than the conflict and compromises of interests themselves . . . thus rewarding inside connections while favoring the strong at the expense of the weak." Although the timber industry regularly highlighted jobs as justifying ensured access to supply, ensuring timber harvests through zoning guaranteed none of the social benefits advocated by MacKaye and others at the beginning of the twentieth century. Despite claims to the contrary, arguments between national environmental and timber interests have, for the most part, remained narrow, advancing self-interest and lacking needed integrative or inclusive thinking.

Integrative thinking in community-based forestry rejects dichotomous approaches and instead calls for a systemwide and multiscale perspective that avoids harsh tradeoffs whenever possible while actively pursuing multipolar, integrative possibilities. It seeks to simultaneously promote sustainable and restorative forms of natural resource management and enhance and improve the quality of life of people and communities linked to the forest. Here we review the three "legs" of the multipolar community forestry "stool," followed by a discussion of a capital assets framework.

Environment

The focus on environment reflects community forestry practitioners' commitment to sustaining forest health, ecosystem function, and biodiversity. Unlike more narrow environmental interests in which ecosystem health is achieved by restricting or eliminating human activities in the forest, practitioners recognize that humans are part of the system and that concerns about health, function, and biodiversity will be addressed through a combination of wild and working landscapes. For many practitioners, wilderness is important but alone is insufficient to meet diverse ecosystem objectives. The essence of working landscapes is the practice of stewardship, the goals of which are to restore and manage ecosystems in a manner that conserves or enhances ecological processes and productivity and generates sustainable streams of ecosystem goods and services. Whether the threat to forest ecosystem sustainability is fragmentation, fuels buildup, forest structure simplification, excessive timber harvesting, conversion, or the negative aquatic and terrestrial cumulative impacts of past land use practices, community forestry groups, almost without exception, are motivated by common desires to reduce or eliminate threats to forest ecosystem sustainability and institute sustainable resource management regimes. These emerging regimes of forest ecosystem management are part of an effort to uncover and understand the diverse place-specific social and ecological processes through whose interaction the forest was created. Understanding these processes is necessary for forest restoration, the attempt to bring forest ecosystem structure, function, and productivity back within historic ranges of variability. This is a primary initial focus of many community forestry efforts. It represents an implicit critique of dominant modes of forest management (and of the models of forest research and science that support them) that emphasize centralized systems of government control and order and use blunt scientific and political categories of forest type (Romm 2000).

Economy

Economy is central to the success of any community forestry effort. The failure of the market to require investment and the general failure of the beneficiaries of forest outputs and products to invest in sustaining ecosystems and communities have contributed to the impoverishment of both. Many financial structures, such as the nature and structure of publicly held companies controlling forestland, inheritance and ad valorem tax laws, and high-interest bonds, encourage the liquidation of natural capital rather than its stewardship. Company takeovers and the increasing concentration of the wood product industry that took place in the late 1980s and 1990s accelerated forest liquidation to pay off acquisition debt. The growth rate of forests rarely, if ever, approaches company shareholders' desired rates of return, making it extremely difficult for publicly held companies to justify long-term management of complex forests, especially forests with late-successional characteristics and old trees. Forest ecosystems provide many important products and services, such as water, biodiversity, and carbon sequestration, for which there are no market mechanisms for recognizing and capturing their value. And there are few financial incentives that encourage and promote investment in ecosystem maintenance and restoration. The result is long-term disinvestment in ecosystems and communities.

These barriers to sustainable forms of ecosystem investment and strategies for overcoming them are further discussed in Chapter 9. A key characteristic of these strategies is the emphasis on advancing mechanisms in which prices reflect the full suite of forest ecosystem services and products and strengthening possibilities for activities such as value-added local processing that increases the flow of economic benefits from forest management to workers, local communities, and local forest ecosystems.

Globalization of the timber and nontimber forest product trade offers other challenges and opportunities for community forestry. Foreign demand for high-quality matsutake and other mushrooms fuels the collection of these special forest products throughout the Pacific Northwest. Challenges include conflicts over access to and control of the resource, sustainability of harvest levels, and the distribution of profits among the different actors involved in the mushroom collecting, buying, processing, and marketing.[2] Log exports, like those to Japan from the Pacific Northwest, from New Zealand to the Pacific West, and to Canada from New England, significantly reduce potential for local value-added processing. Loggers and contractors from Canada who enjoy national healthcare and other social benefits easily underbid their domestic U.S. counterparts in New England. The forest certification movement holds the promise of harnessing market forces to recognize and compensate landowners who practice socially and ecologically responsible forest management, but, to date, this has not been achieved. Thus, although global trade tends to increase the distance between point of origination and point of consumption and sometimes undercuts community forestry efforts, in other contexts it may provide opportunities to generate financial capital that may be reinvested back into natural capital and communities.

Equity

Equity is fundamentally concerned with the assertion of diverse claims, local and nonlocal, to forests and the resources they provide. Equity is itself a multidimensional concept that refers to the distribution of power, knowledge, economic benefit, and, overall, an expansion of human freedom. Nobel Prize winner Amartya Sen (1999:53) describes the expansion of human freedom as "both the main object and the primary means of development." Sustainable forest-related development incorporating equity concerns therefore must confront the issues of who is included and who benefits in terms of distribution of freedoms. Community forestry practitioners are reasserting claims as diverse as the validity and valorization of local knowledge and the rights of local communities and those who work in the forest to steward themselves and the forests that support them. Addressing and resolving these claims and others involves a complex renegotiation and realignment of the interests and claims of local and nonlocal groups. Whereas the dimensions of environment and economy involve primary reliance on the state and the market, respectively, equity calls for local residents and workers to engage with both, and with a particular focus on advancing institutions to promote modifications to the political, legal, and economic structures that have governed forest resource management.

Through the wide-ranging discussions that took place during the workshops and from conversations with community-based forestry leaders from around the country, it became clear that many communities—Latino, African American, Southeast Asian, Native American, and poor white—have yet to participate in meaningful ways in the community forestry movement. The reasons for the lack of participation of these groups are complex and varied, and we focus on them more directly in Chapter 6. But if community forestry is to be an inclusive movement, if it is to fully embrace its own self-avowed goals of participatory democracy and democratic renewal, then the groups, organizations, nonprofits, agencies, and foundations that practice and support community-based forestry will have to seek out and actively and genuinely engage the diversity of groups and communities that rightfully are a part of it.

Fostering the development of independent, higher-capacity community-based organizations and developing more inclusive processes involve common challenges such as improving communication between groups with different languages, cultures, traditions of participation and leadership, and histories. They also involve the challenges associated with working with groups that sometimes exist on the political and economic margins of society, groups that may be invisible in society, and groups that have had little experience of the positive, empowering potential of democratic political institutions. Therefore, including equity means genuinely engaging these groups and tackling the knotty, time-consuming, and challenging task of building capacity to enable participation. This challenge is compounded by the fact that in some cases communities of color and poorer white communities compete with each other for access to forest resources and employment.

Equity is also an integrative process in itself, advancing work that simultaneously promotes sustainable and restorative forms of natural resource management and enhances and improves the quality of life of people and communities whose well-being is linked to the forest. Because the causes of and solutions for environmental and community degradation are interlinked, only an integrative and inclusive response to the challenges of environmental and social degradation will restore ecosystems and revitalize the people and communities that depend on them.

Community Scale and a Capital Assets Framework

Community forestry involves an approach to resource management that integrates and advances environment, economy, and equity together in place.

This approach is unique not only because of its multidimensional focus but because of its focus on place and community-scale solutions. Interest groups focus on economy and the environment at scales far larger than communities and typically ignore local integrative work. Progressive Era–inspired science and policy operate at large scales as well and advance command-and-control approaches that are unresponsive to local variation and local integrative work. In sharp contrast, community-scale work is a bottom-up approach to problem solving, involving local people working on integrative solutions at the appropriate scale. What is unique about this approach is the recognition that successful problem resolution is effective at the community scale. Because of the resistance to bottom-up approaches, community-scale work requires considerable community capacity for success.

The capital assets framework facilitates understanding of the components that are both necessary for and advanced by bottom-up work and community development. This assets framework also clarifies the separation between communities and forest resources and the institutional barriers that prevent their reconnection. The assets framework builds on and extends work in which ecologically minded economists view natural capital as essential to human health and not, as conventional economists treat it, substitutable with financial capital or labor (Prugh 1995); poverty alleviation is achieved by building financial and physical capital (Sherraden 1991) and natural capital (Boyce and Shelly, forthcoming); and forest community well-being is reconceptualized (Kusel 1996, 2001). The assets approach involves a focus on system assets or stocks of natural wealth and community wealth that are integral to community and worker well-being. Capital assets offer a framework to enable deeper probing of the community–environment relationship and a structure for addressing economy, environment, and equity issues.

A fundamental premise of community forestry practitioners is that the natural capital of the forest and community well-being are interdependent. Natural capital is made up of diverse biophysical elements of a forest ecosystem (Schumacher 1973; Bunker 1985); elements include soil, forest structure, the related aquatic and riparian ecosystem, and biodiversity. Community well-being consists of five capital assets: *physical capital,* which includes a community physical infrastructure (e.g., sewer systems, business parks, capital assets such as equipment, housing stock, and schools); *financial capital,* which includes money, credit, and other financial resources available for local use; *human capital,* which includes the skills, education, experiences, and general abilities of members; *cultural capital,* the myths, beliefs, norms, and lifeways that organize groups and facilitate survival; and *social capital,* which includes the willingness of residents to work together toward community goals (and not just self-interest goals). These five capitals together make up community capacity, which is how community members collectively respond to stresses, create and take advantage of opportunities, and meet member needs. Indeed, a great deal of community-based forestry work has focused on attempting to reestablish the long-broken natural and community capital linkage. This same linkage is one that MacKaye and others attempted to advance at the turn of the twentieth century rather than the "trickle-down" idea of development that prevailed, in which natural capital extraction (timber harvest) yields plentiful and sustainable jobs that, in turn, lead to community well-being. The recently growing national and global competition in the wood product industry and a thousand mill closures in the Pacific West in the last half of the twentieth century have made clear the fallacy of this thinking.[3]

Figure 5.1 shows natural capital and community capital separated by a dark line that represents institutions that mediate relationships between the two. Community is inclusive of both place-based and worker (including mobile worker) communities. Communities and workers may overlap, but because the majority of woods workers in the West and the South today are mobile and because the issues they face are different and often do not overlap with place-based issues, communities and workers also must be viewed as separable. Bellah et al. (1991:289) devote a book to discussing how institutions mediate the relations between humans and how we need to think creatively to solve many of our problems. Like the framework presented here, they point out that institutions mediate our relations to the natural world and,

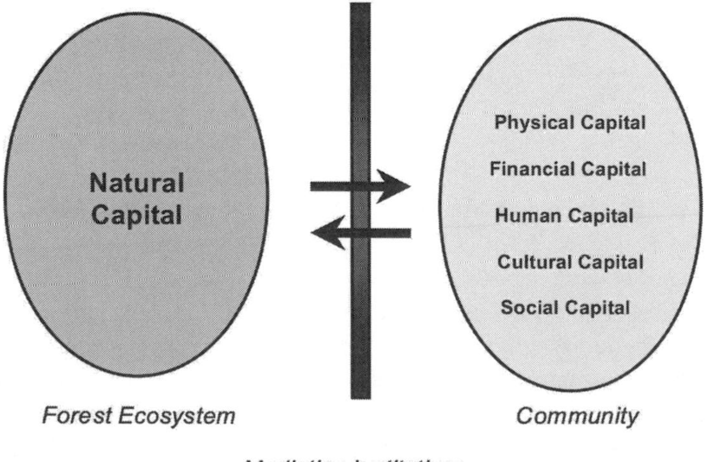

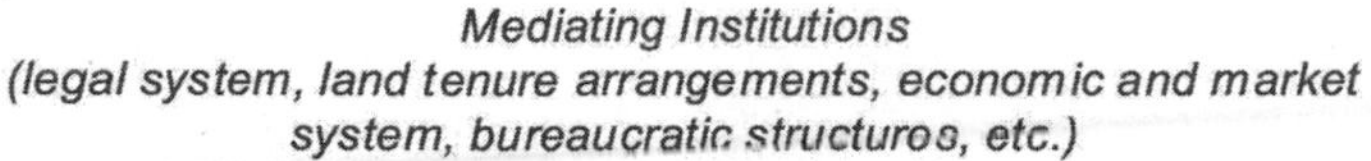

FIGURE 5.1 Linkages and Barriers: Natural and Community Capital

recognizing the issue that blinded Pinchot (i.e., our propensity to seek technical solutions instead of focusing on institutions), they state, "The American tendency to think that social and institutional problems are basically technical is related to the assumption that the central value question is already settled: that institutions are there to serve the private ends of individuals. Yet we have seen that this leads to contradictions and conundrums, and obscures awareness of the many destructive consequences of our current institutional patterns."

To highlight one all-too-common example of a capital depleting role of institutions, natural capital may be converted to financial capital that passes not through but by a community entirely as the institutions of private property and the market allow timber purchase, harvest, and movement elsewhere for milling by labor from outside of an area. Labor from the local community doesn't benefit, and mobile laborers who are hired may be paid substandard wages, are offered no opportunities for learning or advancement, and may work in unsafe conditions. Financial capital obtained from the sale of wood, beyond that which is paid to workers and spent locally, flows to (typically) out-of-area company shareholders. From a community-of-place perspective, if the harvest removes most of the biomass, the community loses out with the harvest of trees in which aesthetic values inhered and loses further if the forest soils, dependent species, and watershed are compromised by timber harvest. In this example, the conversion of natural capital to financial capital is anything but benign as the process of conversion impoverishes the land, the community of place, and the worker community. Or, alternatively, environmental, economic, and equity concerns are ignored at the scale in which community and people work and live as financial capital is generated and contributes, in the aggregate, to the gross domestic product. Private property and the market are only two of the many different kinds of institutions that may lead to disinvestment and local impoverishment, but because this kind of relationship is not inevitable—because there are places where investment using these mechanisms occurs—they must also be seen as holding the potential to contribute to asset building and investment in natural and community capital. Unfortunately, however, depletion of natural capital and impoverishment of worker and place-based communities have historically been more the norm than the exception.

Failure to recognize the role of institutions that mediate the relationship between community capitals and natural capital has contributed to a misunderstanding of the causal factors of diminished natural and community capital and a misunderstanding of the mechanisms needed for investing in and improving both. This failure contributes to the unproductive "owls versus jobs" kind of arguments. Indeed, it is in response to the impoverishment of communities and landscape and the failure of institutions to recognize the importance of community-scale work that a number of community-based approaches have arisen. Understanding and working to improve the performance of institutions to reconnect management of natural capital with communities of place and workers, whether it is through the market, governance structures, or a system of science responsive at multiple scales and to multiple interests, will ultimately lead to better environmental stewardship, improved economy through capital investment, and equity through the recognition of the multiple communities and corresponding capital assets dependent on sound resource stewardship and improved democratic practice.

Conclusions

The environment and economy legs of the community forestry stool have been the primary focus of attention by community forestry nonprofits, national-level community forestry organizations and networks, and, to a lesser extent, government agencies and national policy makers—in short, the people and institutions involved in what may be called the formal community forestry movement. Ethnic and racial minority communities and also poor white communities, whose livelihood and well-being depend on the forest, have yet to fully participate in the formal community forestry movement, even though they may practice community forestry. These communties include black farmers in the Deep South who own or owned forested areas, white woodlot and farmland owners in the Midwest, Latino forest workers in the Deep South and Pacific Northwest, white Appalachian woodlot owners, Latino special forest product gatherers in the Pacific Northwest, Native American forest managers and gatherers across the country, Cambodian, Lao, Mien, Hmong, and other Southeast Asian nontimber forest product gatherers in the Pacific Northwest and increasingly the Great Lakes region, mobile forest workers (mostly Latino), and other loggers and contractors.

Although people of color and blue-collar workers have suffered disproportionately, draining natural capital from the land without reinvesting in the forest or in communities has separated forest communities across the country from the forest and has degraded land and communities. The lack of institutional mechanisms to ensure forest or community investment—and more meaningful community–forest connections—has become a focus of the community forestry movement. Developing wider acceptance of community-scale solutions and integrating the equity leg of the community forestry stool with a focus on environment and economy are two of the major challenges for community forestry practitioners.

Yet progress at multiple levels has also been made regarding important aspects of equity. Some examples include the following:

Public lands management agencies are embracing the principles of collaborative ecosystem management at the national level in administrative language and policy codes and at the local level through collaborative community-based natural resource management planning processes throughout the western United States (Kenney et al. 2000).

Changes in Forest Service contracting procedures, such as "best-value" contracting and packaging some contracts into small sizes, further demonstrate agency changes that promote community forestry.

The Secure Rural Schools and Community Self-Determination Act of 2000 (S. 1608, the "county payments bill") embodies and reinforces the principle that counties with large areas of public lands within their boundaries deserve monetary compensation from the federal government to balance benefits nonlocal communities receive through public ownership against forgone tax revenues for local government that public ownership represents.

Legislation currently under discussion in West Virginia would radically alter the current ad valorem tax on forestland to reduce the negative incentive of the current tax structure for forest stewardship.

State and federal forest worker retraining programs, particularly in the Pacific Northwest, represent equity-based attempts to redistribute the costs and benefits of changes in the forestry sector.

The landscape of community forestry takes on dramatic and dynamic features when the multiethnic, multiracial, and multicultural dimensions of the relationships between communities and forests and asset improvement are considered. The importance of understanding this landscape, the extent to which the formal community forestry movement has yet to engage with this more diverse landscape, and the implications of engaging with this landscape to achieve goals of community forestry such as democratization and civic participation are key themes that will be addressed in later chapters.

CHAPTER 6

Democratic Renewal and Revival

> Only if oppressed groups are able to express their interests and experience in public on an equal basis with other groups can group domination through formally equal processes of participation be avoided.
>
> —*Young (1990:95)*

Most workshop participants and many of those we interviewed described community forestry as a way to strengthen participatory processes through face-to-face deliberations about the management and utilization of natural resources.[1] Seen in this light, community forestry empowers people and communities to take a more active role in the management, processing, and use of forest and other resources through the devolution of management and decision-making authority to local levels. The espousal of more direct, participatory institutional structures is associated with a broader threefold critique of the dominant model of political relations and practice (interest group pluralism), the organization and practice of science, and the inability of markets to recognize and value many nontimber forest products and ecosystem services and to generate the revenue streams necessary to sustain diverse forest structures, forest restoration, and community well-being. These three themes are the focus of this chapter and Chapters 8 and 9, respectively. This chapter discusses the potential of community forestry to strengthen democratic participatory structures and processes. The chapter begins with a critique of interest group pluralism from the perspectives of participation, community well-being, ecological health, and equity. This leads to a discussion of the potential of community forestry to advance participatory democracy through the sorts of community-based deliberations and engagement associated with civic republicanism, the citizen-based, face-to-face forms of public deliberation and grassroots political discourse that Thomas Jefferson championed. After this we discuss cautions against the wholesale embrace of civic republicanism. After a brief review of the philosophical and pragmatic reasons why equity and justice must be a primary objective of community forestry, the chapter addresses some of the ways the movement is grappling with equity issues. This leads to a discussion of strategies for promoting equity within community forestry. The final sections of the chapter discuss the linkages between local empowerment, participatory democracy, and community forestry.

A Critique of Interest Group Pluralism

Those who promote community-based forestry are quite explicit about the tension between the forms of social interaction, decision-making processes, and collaborative structures for which they advocate and the dominant model of political relations and practice. In interviews and workshops people repeatedly stated that community-based forestry represents a challenge to the dominant conflict-based political system and to the "conflict industry," a term that captures how, in representative democracies, individuals organize into interest groups to collectively advance their interests and defend them against others with competing values and interests, which, if taken to its logical extreme, can result in conflict for conflict's sake and policy gridlock. This form of representative democracy is known as interest group liberalism or interest group pluralism.[2] To avoid confusing the ideological meaning of *liberal* with the political theory of liberalism, which is our focus, the term "interest group pluralism" is used.

The basic principles of interest group pluralism were set forth in the U.S. Constitution. One of the more controversial questions those who drafted the Constitution considered was the degree to which the American political system was to be a representative or a direct, participatory democracy. James Madison, fearful of the "tyranny of the majority," argued forcefully for a representative form of government. Thomas Jefferson championed a more participatory and deliberative model of political process. Whereas the latter was more populist and provided for direct citizen participation in politics through local face-to-face discussions and deliberation, the former sought to insulate politics from the populace through establishment of representative rather than direct political structures and processes. In the end the framers of the Constitution chose a representative democracy.

In a representative democracy people are assumed to be strongly individualistic with set values and preferences. People pursue their interests by joining an interest group that promises to advance an agenda consistent with their values. Different interest groups compete with each other for people's loyalties; those that accrue the most members and money are best able to influence legislation, the development of regulations, and the flow of government benefits and tax revenues. Although different interest groups often vie for political influence and resources, they may also create alliances and bargain with each other to mutual advantage. "Winning depends on getting others on your side, making trades and alliances with others, and making effective strategic calculations about how and to whom to make your claims" (Young 1990:72). Government policy results from this process of competition and bargaining between interest groups.

When examined from the perspective of community forestry, at least four weaknesses of interest group pluralism become apparent. First, the practices and institutions of interest group pluralism are not participatory; if advancing a perspective is a contest between numbers and funding, organized national interests will always prevail over unorganized communities and people and community-scale solutions. Second, interest group pluralism institutionalizes and reinforces self-interest as the basis for political engagement. This militates against recognizing and developing common values and integrating diverse interests and perspectives. Third, interest group pluralism is incapable of responding to ecological problems, which tend to be characterized by high degrees of uncertainty and complexity and necessitate approaches that incorporate feedback, coordination across problems and actors, and resilience and flexibility (Dryzek 1998:586). None of these attributes fit comfortably in the structural confines of a pluralist democracy. Finally, interest group pluralism generally does not address issues of equity, fairness, and justice. Instead, it tends to reinforce the unequal distribution of resources, power, and privilege because advantaged groups are able to use their position to influence the policy-making process in ways that preserve their advantage. Thus, interest group pluralism tends to exclude local communities and their ecological knowledge as well as historically disenfranchised groups such as forest workers, especially unorganized people of color. It reinforces existing inequalities in the distribution of resources and power, generally precludes the application of community-based knowledge in environmental management, and does not address social inequality in a systematic fashion.[3] The remainder of this section expands on each of these points.

Lack of Participation

Participation is an important condition and element of justice within a democratic system. Unfortunately, interest group pluralism measures up poorly with regard to it. Direct participation of individual citizens generally is not possible because of the dominance of well-organized interest groups. These groups promote their own interests through government channels that are rarely subject to public discussion and debate. Major policy decisions often are made in private, through complex, informal negotiations involving government agencies and the interest groups with enough power to access policy-making arenas. Individual citizens are prevented from participating in the policy-making process and may be kept in the dark regarding the nature of the proposals discussed and how decisions were made. Furthermore, there are few if any public forums for discussing the overall distributional effects of these processes and the institutional rules, practices, and social relations that produced them.

One-size-fits-all policy making that ignores local variation and excludes communities and rural people from forums in which decisions that directly affect them are made has motivated rural communities to organize, especially in the West. Although industry and environmental groups have well-developed, organized, and powerful interest groups and lobbying capabilities, few rural communities have the equivalent access and political clout to influence policy making. And many forest workers, especially mobile or migrant workers, tend to be even more disenfranchised from the policy- and decision-making arenas that affect them. Again and again, policy, legislation, and administrative decisions have been drafted and approved with little or no thought to their impact on rural communities or workers, nor with the kinds of consultation that effectively engages these groups. In many interviews and during workshops, participants discussed the importance of strengthening the voice and presence of communities and workers at both state and national levels.

Self-Interest Versus Common Interest

Interest group pluralism is predicated on the assumption that people's values are set and their views on issues predetermined. The practices and institutions of interest group pluralism reinforce this perspective; interest groups, comprising like-minded people, are designed to strategically use their access to political and policy-making arenas to forward the agendas and interests of their constituents. This process occurs in a context of competition with other interest groups whose constituencies have coalesced around competing values and views. Structurally, this process almost inevitably leads to a lack of trust between various interest groups. In an interview, Mark Rey, currently Under Secretary for Natural Resources and Environment in the Department of Agriculture, stated that trust between environment and industry groups was "almost nonexistent" at the national level. He argued that the structural characteristics of the political process have led to organizations that are unequipped and unable to communicate. These groups are "openly hostile" to long-term dialogue; instead, they focus on achieving and managing short-term gains consistent with the 3- to 5-year election cycle. It is no wonder that this process has led to entrenched conflict and gridlock.

The distrust that results from interest group pluralism can also breed cynicism. Thus normative claims that a group represents a common or just interest are simply seen as strategic rhetoric, a calculated decision to win in the next round. Many observers have noted the ways in which the structure of the dominant political process militates against the development of common interests and concerns. One of them, Val Plumwood (1998:573), political philosopher and environmental historian, notes in this regard that "the liberal interest group model which treats people as private political consumers provides little encouragement for the development of any public ecological morality, for collective responsibility or problem solving, or for people to transform their conception of their interests, their convictions or sympathies in response to social dialogue with affected groups." This feature of pluralism perhaps explains why community forestry's attempts to identify common interests and to develop shared values through face-to-face interaction, have threatened interest groups.

Difficulty of Conserving Ecological Values

Perhaps not surprisingly, interest group pluralism also measures up poorly when it comes to valuing and prioritizing ecological considerations. As has been discussed in prior chapters, the capitalist market system in conjunction with political structures based on interest group pluralism have neither maintained nor enhanced the ecological productivity of forests nor supported the development of sustainable rural economies and high-capacity rural communities. Indeed, it is the drawdown of natural capital, in combination with the disenfranchisement of rural communities and workers and concomitant environmental gridlock, that led to the emergence of community forestry in the western United States. Although sole responsibility for forest degradation and associated environmental concerns cannot be attributed to the dominant model of political relations in this country, it is commonly recognized that interest group pluralism has led to resource extraction for short-term monetary and political gain and that these gains have compromised environmental stewardship objectives.

Inattention to Equity

Finally, interest group pluralism does not adequately address issues of equity, nor does it satisfactorily provide mechanisms for enfranchising and empowering marginalized or disempowered people, groups, and communities. On the contrary, the practices and institutional structures of interest group pluralism have tended to reinforce the existing inequitable distribution of resources, power, and privilege within society. Interest group pluralism provides few sources of leverage for disempowered groups to challenge the status quo and assert a more equitable distribution of resources and access to economic opportunities, decision-making arenas, and the policy process.

The links between the social inequality and marginalization associated with interest group pluralism and the system's inability to prioritize ecological values have been discussed in other contexts. For example, contrary to the oft-quoted phrase "Poverty is hierarchical, while smog is democratic" (Beck 1995:60), Plumwood argues that the conditions, experience, and effects of both poverty and ecological degradation are systematically and unequally distributed socially and spatially within interest group pluralism. Using the concept of "remoteness" she shows how privileged social groups are able to distance themselves from the immediate consequences of ecological degradation, such as forest or fishery depletion and toxic contamination, whereas less privileged groups, including resource-dependent communities, are not. Because of the links between social inequality and ecological damage, these issues can be effectively addressed only in a coordinated fashion. In the following passage Plumwood (1998:573) suggests why:

> My argument implies not only that the inegalitarian power structure of liberalism is ecologically irrational, but also that the political and communicative empowerment of those least remote from ecological harms must form an important part of strategies for ecological rationality. There are many specific contextual forms this empowerment might take, such as access for community action groups to resources like public funding, but its general conditions surely require institutions which encourage speech and action from below and deep forms of democracy where communicative and redistributive equality flourish across a range of social spheres.

Plumwood's calls for local empowerment resonate strongly with arguments community forestry practitioners set forth regarding the importance of strengthening local voice and decision-making capacities as a necessary but insufficient condition for simultaneously and effectively addressing social and ecological degradation. However, community forestry practitioners and advocates take Plumwood's analysis a step further. Whereas Plumwood acknowledges the links between marginalization and inability to preserve ecological values within the context of interest group pluralism, advocates of community forestry argue that local empowerment is also necessary because it is at the community scale that important forms of knowledge regarding forest management exist. Equally importantly, it is only through the sorts of face-to-face interaction, communication, and deliberation that occur at the local and regional level that seemingly intractable conflicts can be resolved as, through deliberative processes, trust slowly builds, values and priorities shift, and the radical center associated with community forestry emerges. Thus, as the next section demonstrates, community forestry appears to contain at least the seeds of promising alternative processes and institutions that address the weaknesses of interest group pluralism.

Promises of Community Forestry and Participatory Democracy

Community forestry proposes an alternative vision of politics from that entrained in interest group pluralism. In many respects this alternative vision addresses the shortcomings of interest group pluralism, particularly the lack of participation, the entrenched conflict between interest groups, and the difficulty of adequately addressing ecological issues. Community forestry practitioners call for increased community participation in natural resource planning and management on public and private lands, for more empowering and collaborative planning processes than the "traditional" forms of public input, and for increased consideration of the local effects—social, economic, and ecological—of management decisions. Many community forestry groups emerged out of intense frustration with the social, economic, and ecological effects of practices associated with interest group pluralism. People interviewed for this research project traced the origins of community forestry to attempts to loosen the gridlock and entrenched natural resource conflicts so pervasive on public and private lands in the West. In most cases the gridlock resulted from escalating conflict between nonlocal interest groups pursuing their own agendas using the logic and strategies associated with interest group pluralism. While policy and management responses to forest and ecological degradation were developed, people and communities were not empowered to participate in the policy-making process; as a result, the proposed solutions almost invariably left them out. The continuing social, economic, and ecological toll on rural communities eventually led to the spontaneous ignition of collaborative community-led grassroots groups across the country.

As discussed in Chapter 5, the rapidly growing numbers of groups seeking to open up local decision-making space between the confines of government and the press of national interest groups did so by attempting to expand the sphere of civil society, the domain that lies between individuals and family and state and corporate institutions. To use Plumwood's terminology, most of these groups seek to reduce "ecological remoteness" by empowering communities affected by ecological degradation to actually share their views, be heard, and participate in the development of effective and acceptable solutions to ecological degradation that simultaneously strengthen community capacity and well-being.

A common argument community forestry practitioners use to justify why they might have unique insights into effective ways to restore and manage forest ecosystems is that the nature of the challenges associated with forest restoration requires the kind of long-term stewardship, awareness of site-specific conditions (historical as well as current), and ability to adopt an adaptive learning management philosophy that groups situated close to the forest generally have. This argument is supported by scholars such as political scientist John Dryzek, who have described appropriate political structures for addressing today's pressing environmental concerns. Dryzek (1998:586) suggests that ecological problems are characterized by complexity and uncertainty as well as collective action issues. He argues that an appropriate political structure for addressing these challenges includes several attributes. It incorporates feedback, coordinates across different problems and actors, and responds to changing conditions (flexibility) and sharp perturbations or disequilibria (resilience). Dryzek argues that although democracy is the best-suited political system for achieving political structures that approximate these attributes, interest group pluralism is an ill-suited variation for this purpose. He suggests that political structures with the characteristics necessary to address the complexity, uncertainty, and collective action nature of ecological problems must have a strong deliberative element.[4] This argument leads him to conclude that the public sphere or civil society constitutes a deliberative democratic model capable of performing the coordinating functions necessary to address ecological challenges. He argues that groups (often self-organized) within civil society that emphasize deliberative engagement with the full diversity of stakeholders, that have egalitarian internal politics, and that generally subscribe to consensual decision-making processes are better able to respond to the challenges associated with ecological problems than the hierarchical bureaucracies and profit-oriented goals associated with the practices and institutions of interest-based pluralism. In many ways, the characteristics of community forestry groups conform to his description of groups within civil society that may be able to respond effectively to ecological challenges.

Practitioners, advocates, promoters, and students of community forestry explicitly acknowledge that community forestry involves fostering participatory and deliberative forms of democratic process, in contrast to interest group pluralism. For example, author Daniel Kemmis provides an in-depth discussion of the rise of grassroots-based collaborative resource management efforts in the West. Rooted in watershed and Jeffersonian-based forms of democracy, Kemmis argues that these groups and the pragmatic collaborative processes they espouse constitute a decentralization of sovereignty and a reinvigoration of federalism. Kemmis (2001:181) argues that the "collaborative movement" that is sweeping across the western states harkens back to John Wesley Powell's calls for watershed forms of political organization and local institutions based on cooperation. Furthermore, it represents a more appropriate future direction for western land management than the "prevailing centralized and adversarial decision-making structures and . . . the region's arbitrarily bounded political jurisdictions."

This interest in pursuing participatory democracy arises from the experience and awareness of the weaknesses of interest group pluralism when viewed from the perspective of disempowered communities and ecological rationality. Many community forestry advocates and practitioners appeal to the possibility of developing a shared understanding of common interests across diverse social groups, enabled by the sorts of direct face-to-face deliberative processes that Dryzek describes. This approach places faith in the ability of people, through dialogue and civic engagement with others in their community, to participate in developing mutually acceptable policy directions for natural resource planning and management. Rather than a winner-take-all pitched battle between opposing interest groups, community forestry advocates espouse a "republican" mode of political discourse that champions local empowerment and participation and provides incentives for individuals to rise above their particular interests and concerns in an effort to identify and then take steps toward realizing the "public or common good." In this vein Goergen et al. (1997:11) note that "most of the success stories in community decision making involve republicanism," and they discuss the implications of local democracy for the role of resource professionals such as foresters and for large-scale bureaucracies in general. Rather than the "liberal" image of the professional forester "armed with specialized training and backed by agency mandates, [assuming] control in order to provide effective management" (Wellman and Tipple 1990:76, cited in Goergen et al. 1997:11), the republican approach, grounded in participatory democracy, "would suggest that forestry professionals share knowledge with citizens and represent the broad public interest—that they become a voice for those who cannot be at the table as well as for the resource itself." The implications of this view of the role of the professional for the practice of science, and in particular civic science, are discussed further in Chapter 8.

A key tenet of civic republicanism is that through open discussion and dialogue involving stakeholders and interested parties, a shared vision of the common good will emerge; this assumes that people's values are not predetermined and immutable and that through the deliberative process individuals transform and modify their own values and interests in the process of developing that shared vision. Because civic republicanism involves extensive deliberation to craft shared visions of the public interest, it entails nonhierarchical organizational structures that facilitate communication about knowledge and values across a diverse and broad spectrum of stakeholders.[5]

Does Civic Republicanism Promote Equity and Justice?

The images associated with civic republicanism have an inherent appeal: citizens gathered together in a public meeting place to engage in face-to-face deliberations over issues of common concern, creating community and building trust by shaping a mutually acceptable vision of society, economy, and ecology. The image includes professionals who share insights with the community based on their expertise, and community members themselves contributing their own perspectives based on a long association with the issues at hand. By carving out and using public spaces for such deliberative engagements, civil society is strengthened and overall community capacity grows. Within community forestry, the social processes associated with civic republicanism remedy many of the weaknesses associated with interest group pluralism, such as the exclusion of people and communities, the "conflict industry," and inadequate attention to ecological values and long-term stewardship. And, in fact, this image is congruent with many examples of community forestry around the country.

The images of community forestry process and outcomes that civic republicanism conveys are appealing because they are good. Empowering communities historically "whipsawed" by political processes to which they had little or no access, loosening entrenched conflict, crafting shared visions of long-term forest ecosystem stewardship that simultaneously support desired ecological, social, and economic values, building trust and broader community capacity, and democratizing science and bureaucracy—these are all positive steps that justifiably enjoy broad appeal across diverse sectors of society. These attributes account for the attraction of community forestry to philanthropic organizations, nonprofit organizations concerned about social and ecological sustainability, and many in the public sector. The positive vision of community forestry also goes a long way in explaining the willingness of community members (as well as public servants and others) to volunteer countless hours and endless energy to make the community forestry vision a reality.

Do these justifiably appealing images of community forestry, conveyed through the lens of civic republicanism, incorporate concerns about social equity and justice that extend to all groups that use, work in, and depend on the forest resource? Is there a place for everyone with a stake in the forest to occupy public spaces for deliberating issues of common concern? The short answer to this question is no; civic republicanism does not necessarily incorporate equity and social justice concerns. As Foster (2002:150) notes, "there is nothing inherent in collaboration . . . that promises a resolution to the problems of unequal representation and influence that underlie conventional decision-making processes." The rest of this section, drawn from the general literature on civic republicanism, shows why.

The Paradox of Democracy

The term "paradox of democracy" refers to democratic processes that are exclusionary and that produce unjust or inequitable outcomes that undermine broader democratic values and goals. Though commonly associated with interest group pluralism, paradoxes of democracy may also emerge from civic republican processes.

The roots of the paradox of democracy go back to the origins of civic republicanism and public deliberation, to Enlightenment thought and the notions of participation that Jefferson embraced. For although the ideal of civic discourse, public deliberation, and republicanism is associated with Jefferson, those who actually were allowed to participate in such discourse were wealthy, slave-owning, white men, and the nature of deliberation was expected to be governed by the rational, scientific paradigm of thought the Enlightenment championed. The public realm, as it evolved in the nineteenth century, came to be associated with "respectable" white men and rational, scientific discourse. Those thought incapable of dispassionate reason, such as African Americans, Native Americans, and women, were excluded from the public realm of citizen discourse. Jefferson and other early American republicans "explicitly justified the restriction of citizenship to white men on the grounds that the unity of the nation depended on homogeneity and dispassionate reason" (Young 1990:111).[6] Although the stark inequalities of the early American version of civic republicanism have since been blunted, exclusionary legacies that suppress difference remain. One of these is the preference for public discourse that is dispassionate, carefully reasoned, and shorn of emotion. The narrow range of acceptable modes of discourse and communication excludes a wide range of other forms and styles of communication and tends to direct the flow of ideas and proposed solutions in directions that may not reflect the full range of values and meanings associated with a particular issue.

In Whose Interest Is the Public Interest?

One purpose of civic republicanism is to provide a process that allows people to craft a mutually acceptable, if not shared, vision through face-to-face dialogue and deliberation. This shared vision, commonly called the common good or the public interest, is arrived at through a deliberative process in which individuals' own values and interests are modified and shaped through dialogue and discussion with others. This central theme of civic republicanism sharply distinguishes it from interest group pluralism, in which the interests and values of individuals are taken for granted, and the state (often, but not always) functions as arbiter between competing interest groups.

The exclusionary history of civic republicanism raises the concern that the "common good" (or as Pinchot phrased it, "the greatest good, to the greatest number, for the longest time" [1947:48]) may be aligned with the interests and values of powerful elite groups rather than reflecting the full diversity of views and perspectives on a set of issues. Civic republicanism is as vulnerable as Progressive Era conservationists were to charges of elitism. This is of particular concern when there is conflict between the interests of those with power and those without it. In these situations consensus-based and participatory democratic processes give more credence to the experiences and perspectives of advantaged groups than to those of disadvantaged or oppressed groups, whose concerns and interests may simply never be expressed or shared. In an analogous manner, management practices and objectives described as "in the public interest" by the scientific professional forestry community may contain disciplinary biases that actually work against the interests of particular groups and communities. Thus, concerns may well arise when it is suggested that "forestry professionals . . . represent the broad public interest—that they become a voice for those who cannot be at the table" (Goergen et al. 1997:11). Rather than asking professionals to speak for "those who cannot be at the table," a more democratic approach would be to undertake the work necessary to accommodate the diversity of voices, perspectives, and communities at the table.

The Ideal of Community?

Advocates of civic republicanism promote communitarian-like visions of the ideal of community as an alternative to the individualistic, self-interested model of political relations and social life associated with interest-based pluralism. The ideal of community privileges face-to-face interaction, often has a strong place-based, localist flavor, and is associated with the positive aspects of civic republicanism discussed in preceding sections of this chapter. In many respects similar to bioregionalism, the ideal of community promotes decentralized, small-scale communities with significant decision-making authority. The human-scale, face-to-face nature of small communities, it is argued, promotes the sorts of direct communication and exchange of knowledge and information necessary for participatory democracy and, in the case of community forestry, advances long-term ecological sustainability.

Despite the appeal of the ideal of community, there are significant concerns associated with the embrace of community as the alternative to liberal democracy. These concerns revolve around the potential for the ideal of community, with its emphasis on common experience and common values, to exclude those with whom local groups do not identify—both those who are spatially proximate to and distant from the community. In this vein, Young (1990:234) notes that "the most serious political consequence of the desire for community . . . is that it often operates to exclude or oppress those experienced as different. Commitment to an ideal of community tends to value and enforce homogeneity." Thus, the ideal of community may reinforce patterns of marginalization and invisibility of the less empowered people within a community. The homogenizing effect of embracing community can mask situations in which the negative consequences of environmental degradation may be unequally distributed because those who bear the greatest burden may be least able to participate in local republican-like democratic processes.[7]

This section has identified several reasons why civic republicanism does not inherently promote equity and social justice. To the contrary, there are strands within civic republicanism, derived from its historical legacy, its understanding of the public interest, and its embrace of the ideal of community, that can perpetuate and reinforce patterns of social and economic inequality. Some of these tendencies, not surprisingly, are visible in the community forestry movement. However, before moving on to address those tendencies, it is useful to deliberate on the question of why equity matters in community forestry.

Why Equity and Social Justice Are Important

There are both philosophical and pragmatic reasons why equity and social justice are important for community forestry. Philosophically, equity, justice, and participation are core social values that have inspired the most transformative social movements in the history of the United States. These values are also central to this country's national identity, self-representation, and presentation in international arenas. Recent scholarship, such as Mutz, Bryner, and Kenney (2002), has begun to draw much-needed attention to equity and environmental justice concerns in the context of natural resource issues. The nexus between community forestry, equity, and social justice may well be the most recent embodiment of successive efforts in this country to redress historical legacies of injustice and to transform social structures that exclude and oppress disenfranchised people and communities. In this vein, Limerick (2002:338) suggests that "many of the dreams of the 1960s and 1970s seem to have migrated to and found a home in the movement known as environmental justice." As a transformative social movement, community forestry shares the lofty goals and aspirations of movements such as the civil rights and women's movements.

In addition to the philosophical and analytical reasons for valuing equity and social justice, there are also pragmatic reasons. Not asking the hard questions raised by an equity and social justice perspective will compromise the movement's success and will eventually undermine it. Thus it is in the interest of all community forestry practitioners, including those already at the table, to be concerned about making the movement as broadly inclusive and participatory as possible. For example, one group with which community forestry practitioners have struggled to engage and reach out to are unorganized forest workers and harvesters, those who labor in the forest and whose long-term livelihood depends directly on sustaining forest ecosystems. Forest workers and harvesters have yet to become full participants in the community forestry movement; their continued lack of participation narrows the movement's support base. Furthermore, a strong worker presence is important to balance the business-oriented community forestry strands. Brown (2001) argues that forest labor plays a pivotal role in the success of community forestry because forest workers have a direct stake in the participatory processes community forestry promotes and because forest workers are the only community-based constituency that has the potential to act as a system of checks-and-balances for the power and influence of business. She notes that environmental groups have correctly identified the potential for business interests to "dominate ecological interests" (2001:296) within community forestry. Indeed, this is one of the main reasons for their resistance or outright opposition to community forestry. Therefore, Brown argues, forest workers must be empowered to participate in participatory community forestry forums because their influence will balance or moderate the short-term profit-maximizing tendencies inherent in business ventures; no other group is structurally positioned to fulfill this role. In this regard Brown (2001:299) notes that "labor's right to participate is not only an essential democratic right and an expression of fundamental justice, it plays a pivotal role creating a reciprocal and accountable system of checks-and-balances in a genuine 'community'–participatory ecosystem management system."

There are other pragmatic reasons why the community forestry movement must take equity and social justice concerns seriously. One good reason is to avoid violent conflict. Almost anywhere in the world, when people are excluded from decision making and prevented from accessing resources to which they consider they have valid rights, struggles over natural resource rights erupt. Depending on the context, such struggles can turn violent and result in forest destruction through arson and other forms of sabotage or property damage. In addition to the high social costs of such civil society breakdowns, forest arson renders moot attempts at sustainable forest management. The importance of this lesson is becoming increasingly apparent in the United States. Violent conflicts occasionally erupt over the access and harvesting of nontimber forest products throughout the Pacific Northwest; the casualties have been social as well as ecological. The fact that the violence most commonly erupts between groups from different ethnic, racial, and cultural (but generally not class) backgrounds makes it all the more important to develop participatory processes that include the full diversity of people involved in the issue. Increasing rates of forest arson in the Deep South and parts of Appalachia and timber theft and associated forest destruction in the Southwest are other examples of the ecological cost of excluding people from management decisions that affect them. The lesson from these examples is that justice and equity are not only lofty philosophical goals but also are central to even the short-term success of community forestry.

Justice and equity are also necessary for the long-term viability of community forestry. One common dynamic that exemplifies this point is the zero-sum framing of relations between local labor and the mobile labor force, in which their respective interests are perceived as competitive rather than complementary. Supporting jobs for local rural communities is an important goal, but in the long run it cannot come at the expense of excluding the mobile workforce from the community forestry movement and the worker benefits it seeks without undermining the movement itself. Although in the short term it may be possible to win contracts that provide employment to local communities (although given the obstacles to arranging such contracts in many Forest Service districts even this has proved difficult), exclusionary forms of contracting are simply not sustainable in the long run.

Unless alliances develop between diverse groups of forest workers, place-based and mobile, for developing and articulating a shared vision of the elements of a high-quality work environment and then demanding that those elements be provided for all forest workers regardless of whether they are local, mobile, or migrant, it will only be a matter of time before the favorable terms and conditions for one group are undermined by programs such as the proposed guest worker legislation that enable corporate forest owners and contractors to draw on cheap supplies of labor from south of the border, with little attention paid to providing a high-quality work environment. The flagrant labor law violations and lack of attention to worker health and safety issues in such situations are well known.[8] Although programs such as the U.S. Forest Service forest stewardship pilot projects and provisions for best-value contracting are promising, they will not withstand the combined force of budgetary pressures and the "race to the bottom" of the corporate world, especially if other government programs such as the guest worker program make available cheaper supplies of forest labor.

On one side, agencies such as the Forest Service, backed by Congress, must provide clear leadership direction to contracting officers that guaranteeing workers a living wage trumps getting the most acres of brush cleared per dollar; on the other side, the community forestry movement must seriously consider how to create strong alliances across the diverse groups that make up the forest labor sector. The task of creating a rights-based approach to forest labor is daunting; it entails nothing short of reclaiming MacKaye's vision of a forest management that sustains vibrant rural communities instead of logging camps or their contemporary equivalent: trailer parks and tent sites. It entails serious strategizing about how to reclaim work and occupation as a basis for meaningful forest enfranchisement. Because of the cross-sectoral nature of this segment of the labor force, in which workers cycle between agricultural, forest, fishery, and service work, the challenge is all the more daunting because meeting it entails addressing workforce issues at a basic structural level. Although achieving such broad-based structural change probably is beyond the scope of the community forestry movement, engaging with allies in forestry (e.g., socially progressive elements of the forest certification movement) and other sectors to address some of the systemic issues related to labor seems to be a useful starting place. Without such a coalition, the current advances that some community forestry groups have made with regard to the terms and conditions of forest work will be short-lived. Sustaining them takes the creation of a broad coalition that includes the full diversity of people who work in the woods.

Even after recognizing the pragmatic value of equity and social justice, which most community forestry practitioners, at least in theory, do acknowledge, it is entirely another matter to actually develop civic republican processes of participatory democracy that are equitable and just. Strategies for advancing such processes are reserved for discussion later in this chapter; the next section addresses some of the current tensions in community forestry, as articulated during the workshops and interviews conducted for this project.

Community Forestry and Participatory Democracy: Some Current Tensions and Challenges

One of the biggest challenges the community forestry movement currently faces is to become as multicultural as those who practice community forestry. Many of the current tensions in community forestry stem from the difficulties associated with engendering the full and meaningful participation of the diverse groups and communities whose livelihoods and well-being depend on the forest but who lack voice in the community forestry movement and in society at large. For the most part, these communities consist of low-income or ethnic minority groups. The barriers to their participation in the civic republican vision of democracy discussed in this chapter are manifold. During interviews and in workshops, two primary thematic areas emerged concerning this issue: concerns that the full diversity of people practicing community forestry are not yet participating in the community forestry movement and concerns that worker participation, especially by mobile and migrant workers, is inadequate.

The Need for Diversity Within Community Forestry

During the Intermountain West workshop in Denver, participants reflected that community forestry is not just for place-based communities and that it has yet to enable the participation of non–place-based groups. It was acknowledged that room must be made for the full expression of cultural, ethnic, class, and worker diversity within forestry. Although many community forestry groups are grappling with these issues, historical patterns of exclusion of marginalized communities are difficult to reverse. These themes, articulated at the Intermountain West workshop, were echoed in many of the interviews. Many interviewees, whites and people of color alike, stated that there remain big challenges with regard to developing inclusive working relations across racial and ethnic divides, especially with respect to non–place-based minority communities. Place-based and migrant or mobile groups self-identify as having been excluded from the table; their shared exclusion and common forest dependence could constitute a strong basis for inclusive rather than exclusive social relations in the community forestry movement. Several people of color argued that community-based forestry is just beginning the process of reaching out to minority communities and that minority communities lack the institutional infrastructure and community capacity they need to participate in community forestry initiatives.

Discussions of the need for the community forestry movement to become more multicultural and ethnically and racially diverse were a central component of the Pacific West workshop. The workshop participants also addressed the serious barriers and challenges to full participation by the diverse groups whose livelihoods depend on forest resources. Several participants pointed out that the people who derive their livelihood from work in the woods, whether in the formal or informal sector or on a permanent or seasonal basis, are extremely diverse. Their cultural, linguistic, class, ethnic, and racial diversity is greater than that of the community forestry movement. The tremendous diversity of this group often is hidden from view because many of these groups are marginalized within the broader society and because place-based communities occupy such a high profile in community forestry. The espousal of civic republicanism and the promotion of the ideal of community within community forestry further mask the profound differences between groups that use forest resources; they mistakenly reinforce the view that minority communities play only an incidental role in forest management and use.

Similarly, civic republicanism can also mask significant class differences between communities. For example, in parts of Appalachia where racial diversity is particularly low, class differences can be acute and of particular relevance for understanding differences between nontimber forest products harvesters and growers. The former tend to come from lower-class groups that own little or no land, whereas growers tend to come from a higher-class position and generally tend to own forestland on which nontimber forest products may be cultivated. This distinction is important for understanding the needs and constraints of these different groups, and it has implications for designing effective outreach and extension programs.

As discussed in the preceding section, the emphasis on localism within the ideal of community tends to encourage the definition of community as a homogenous group, in opposition or contradistinction to those that are different or nonlocal, with whom the community may not identify. In community forestry this definition of community tends to skew the word "locals" to mean white people who have developed customary rights to nearby forest resources and who feel justified in trying to defend "their" forests against "outsiders" (i.e., minority migrant or mobile forest workers or gatherers), even though the forests often are public and the "locals" may live farther way from the forest resource than the "outsiders" themselves.[9] The conflicts that can erupt in these situations can be and have been violent. Resolving these kinds of conflict entails dismantling the institutional mechanisms that preserve oppressive social relations and exclusionary practices. It involves creating in their place institutional mechanisms that promote communication, knowledge sharing, and collective problem solving across diverse groups of people. In places such as the Illinois Valley, Oregon, where the All-Party Mushroom Monitoring Project, an effort of a local community forestry group called the Forestry Action Committee, has reduced tensions between the various collecting communities through an innovative worker-led joint monitoring project and ongoing meetings, it quickly becomes apparent that simply initiating dialogue across linguistic, cultural, ethnic, and other differences is a significant challenge that takes creative and innovative thinking.

Does the "Common Good" Exclude Forest Workers' Interests?

Within the community forestry movement, the discourse of the "common good" may have exclusionary effects. In place-based communities in the West, discussions of how to advance the "common good" (e.g., what is good for the community) often focus on how to encourage small business development, especially using the byproducts of ecosystem management and forest restoration. This model of conservation-based development, which links economic growth with forest restoration, is situated in and subject to the limitations of the dominant capitalist economic and political structures. This model also resonates with domestic and international experiments that encourage and promote small-scale capitalist entrepreneurial activity through microcredit lending programs. Small business and conservation-based development is thought to promote the common good through its combination of ecological restoration work and employment generation. As one hopeful supporter of small-scale business enterprises using forest restoration byproducts stated, through business development and job creation, labor and worker issues will resolve themselves. Consistent with the small business development perspective on conservation-based development, Brendler and Carey (1998:22) suggest that community forestry is spearheaded by nonprofit organizations and "forest-based rural development practitioners" who "believe that forest protection and economic development are inseparable goals." They go on to emphasize the economic development view of community forestry: "Most practitioners are nonprofit organizations. Some nonprofits are associated with for-profit ventures, and an increasing number operate small businesses. Many combine economic development projects with social service programs and environmental advocacy and education." This model of conservation-based development has strong support from philanthropic foundations. From the point of view of workers and underrepresented groups, however, this model holds the weak promise that they will benefit through trickle-down or "by getting a job"; it sidesteps the pressing issue of how to advance work and occupation as a basis for meaningful forest enfranchisement and the development of a rights-based approach to employment as part of that enfranchisement.

The conservation-based development model of the "common good" in community forestry excludes the values, interests, concerns, and perspectives of workers, especially those of importance to mobile or migrant workers. This point was repeatedly made throughout workshop and interview discussions. Comments such as "The market structure works for society in the short run, but not for workers" and "There is a lack of respect for the forest worker and professional" and discussions of the need "to increase the life expectancy of workers through improved safety, for value-added jobs, not just value-added products, and to make community-based forestry relevant to those who work the land" reflect the low priority that workers feel they have been accorded and the clear awareness that what is good for business (large or small, value-added or not) is not necessarily good for workers.

These sorts of issues challenge the ability of participatory democracy and civic republicanism to enable the full and meaningful participation of all groups that depend on the forest. The homogenizing and difference-suppressing tendencies of some community-based decision-making processes can mask the linguistic, class, ethnic, and cultural diversity of these groups. At the Intermountain West workshop it was acknowledged that communities of place do not represent the interests of the mobile workforce in the region and that Asian, Latino, and other mobile or migrant workers and harvesters are not as engaged with community forestry as "they should be." In fact, in some cases, the homogenizing tendencies of civic republicanism can render some groups of forest workers and harvesters almost invisible, especially if they are also marginalized in the broader political and societal context. Therefore, in many cases a first step involves working to make marginal groups more visible in the community forestry movement. Even after the different groups involved recognize each other as legitimate participants and acknowledge that they all deserve to be involved with planning, management, and use activities, there still remain challenging barriers to engaging with such diverse groups. Some of these barriers include the challenge of communication in multilingual settings, working with diverse culture-specific patterns of leadership and collective decision making, and the historical legacy of oppression and marginalization of some groups, especially in terms of their impacts on people's willingness and capacity to speak out in public forums to express their views.[10]

Broadening community forestry to better reflect the diverse array of groups that work and live in the forest entails acknowledging the centrality of all forest workers within the movement; this has yet to happen. However, in the last few years the voices of some forest workers have been growing louder. This is partly because of their growing collective resources and organizational capacity and the work of groups such as the Jefferson Center, Alliance of Forest Workers and Harvesters, and Federation of Southern Cooperatives. The Healing Harvest Forestry Foundation, in Virginia, is another nonprofit that advances forest worker issues. Part of its mission is to restore and rekindle respect, honor, and value for the skills and knowledge of people who log with horses and mules, whom the foundation calls "biological woodsmen." As the foundation's director puts it, this is part and parcel of "enhancing the environmental ethic by enhancing the dignity dividend." As the honor and dignity associated with forest work increase, so will the inherent motivation to steward forest resources. The attempt to raise the dignity of forest work is a response to the widespread devaluation of physical labor in society, including that of forest workers. This is compounded by a general lack of awareness of the connection between forests and societal consumption of forest products.

The fledgling efforts of forest workers to increase their representation in the community forestry movement are still small in scale. Many categories of forest workers, such as large- and small-scale contractors, organized labor, and mobile migrant workers who work for forest contractors remain outside of community forestry.[11] The forest workers and other underrepresented groups with whom community forestry has begun to engage remain, for the most part, at the periphery of the movement. The community forestry movement has simply not yet been able to address worker and labor issues in a direct and systematic manner. This is partly because doing so involves confronting the deep structural issues that consistently operate to disenfranchise workers and undermine efforts to create a rights-based approach to forest work. Efforts to incorporate workers and to advocate for worker issues within community forestry are further hampered by the broader national context of weakened labor laws such as the Wagner Act, overall weak representation for workers, the difficulty of claiming redress for labor law violations, and new initiatives such as the revival of the guest worker program that will further weaken worker bargaining positions. The lack of legal assets of workers, especially migrant workers, immeasurably increases both the difficulty and the importance of working to enfranchise these groups.

There is a profound difference between setting a place at the community forestry table for worker representation and actually attending to the structural, legal, social, and historical complexities and challenges of enfranchising this diverse group of people. As articulated at the Pacific West workshop, accomplishing the latter requires the realization that the community forestry movement is a small piece of a much larger pie, the rest of which is composed of the many groups and people involved in forest management but not part of the formal movement, their relations to each other, to broader political and economic structures, and to natural resources.

An Integrative Movement That Advances Equity and Social Justice

In its ideal form, community forestry promotes equitable and just decisions, integrates and takes advantage of diverse forms of knowledge, and results in management decisions that sustain or enhance desirable ecological conditions. How can community forestry more effectively achieve the goals of participatory democracy and social justice?

Democratic and participatory processes can ensure that the needs and interests of all people will be expressed and that particular interests and voices will not dominate. As long as all people participate in decisions that affect them and all needs and points of view have been freely expressed, decisions rendered within a participatory and democratic forum will tend to be just. When all who are affected by a particular decision participate in the decision-making process, participants' viewpoints will be called into account and challenged along precepts of general fairness and justice. This helps to introduce standards of justice (as opposed to "just us"), fairness, and equality for judging the relative merits of different policies. This dynamic also provides strong incentives for bringing into the decision-making arena the best available knowledge and information; it maximizes the social knowledge that will be shared and used during decision making. Intrinsically, democracy provides opportunities for citizens and social groups to develop and exercise their own capacities. This includes the ability to think of one's needs in relation to others, interest in considering the relations between other people and social institutions, and the ability to reason and debate persuasively. In this regard Young (1990:92) comments that "the virtues of citizenship are best cultivated through the exercise of citizenship."

Once it is accepted, for both philosophical and pragmatic reasons, that advancing equity and social justice is a desirable objective of community forestry, the question arises how best to achieve that objective. On this count, there is a remarkable convergence between the ideas advocated by political theorists and the proposals of workshop participants and interviewees. Both political theorists and community forestry practitioners and promoters argue that this requires a political process that is self-consciously aware of the paradox of democracy, the homogenizing effects of ideas such as the "public interest," and the oppressive side of the ideal of community; and supports the development of participatory structures that explicitly acknowledge and encourage representation of the diverse perspectives and interests of the people involved in community forestry. Young (1990:157) endorses this politics of difference. She argues that an "egalitarian politics of difference . . . defines difference fluidly and relationally as the product of social processes" rather than the traditional, essentializing meaning of difference that has devalued or excluded people because of their group attributes. Within this formulation, which she calls democratic cultural pluralism, "the good society does not eliminate or transcend group difference. Rather, there is equality among socially and culturally differentiated groups, who mutually respect one another and affirm one another in their differences" (1990:163). This form of democratic cultural pluralism "acknowledges and affirms the public and political significance of social group differences as a means of ensuring the participation and inclusion of everyone in social and political institutions" (1990:67). When community forestry is conceived in this manner, it can then become a vehicle for social justice.

Embracing Diversity

The challenges of fully embracing equity and social justice as part of becoming a participatory democratic movement were common workshop and interview themes. An important starting point involves acknowledging the cultural diversity inherent in the forestry sector; this entails bringing submerged differences into view. Understanding the histories of localities and of different social groups and recognizing the social diversity embedded in our own landscapes and historical patterns of social and ecological change is part of this process. The tremendous diversity of the social and ecological landscape of community forestry and the importance of recognizing, understanding, and validating this diversity constitute community forestry's biggest challenges and opportunities.

Enhancing Participation

Many concrete ways of espousing the politics of difference and avoiding the potential shortcomings of civic republicanism were suggested at the workshops and in interviews. At the heart of these discussions was the realization of just how difficult it is to foster the participation of underrepresented groups in community forestry and how important it is to increase their visibility. This is especially challenging when underrepresented groups work in the informal forestry sector or are only seasonally engaged in forest management and product extraction and processing. Engendering the full and meaningful participation of all stakeholders, especially those previously excluded, involves more than setting a place at the community forestry table and inviting "new" participants to join the discussion. As Brown (2001:300) notes, forest workers probably would not immediately take a seat at the table, if it was offered, because a strategy of avoiding the table has evolved in response to decades of exclusion from natural resource decision-making processes and in response to direct and indirect retaliation by employers when workers did attempt to speak out. This points to the need to strengthen workers' forest enfranchisement so that they can participate in public, participatory forums without fear of retaliation. Encouraging participation follows enfranchisement. It involves thinking strategically about how to facilitate communication and collaborative decision making both within and between stakeholder groups. This entails continuing the advances that have been made with respect to membership diversification within practitioner organizations. It also includes developing minority leadership and building institutional capacity within minority communities. Involvement with regional and national community forestry organizations, as well as local initiatives and collaborative processes, generally requires individuals and organizations to have resources to subsidize their participation. The lack of these resources can constitute a barrier to participation by underrepresented groups in multistakeholder discussions and meetings. In some cases resources taken for granted by others are unavailable to underrepresented groups. These could include the luxury of contributing time to participate in community or government or collaborative decision-making forums or even the transportation necessary to travel to the meeting location. In other situations, there may be no institutional mechanisms through which diverse stakeholders can come together to discuss their common or conflicting interests, or some groups may determine that their participation is not worth the time and energy.

Where "public" meetings have been organized, the style of communication and group dynamic may appear to be neutral when in fact they reflect that of the dominant culture. Addressing this issue entails developing self-awareness of the cultural specificity of these sorts of forums and thinking creatively about ways to facilitate the participation of others who are present but who come from different traditions of public discourse, deliberation, and decision making. Accomplishing this entails understanding and acknowledging the importance of the specific historical experiences of different stakeholder groups, the legacy of those experiences, and how they affect a group's ability and way of engaging in public processes and working with authority figures such as law enforcement officials, managers, and scientists.

Another essential aspect of participation discussed during workshops and interviews was the importance of selecting meeting places where participants feel safe and able to express their views without fear of recrimination and retaliation. In areas characterized by conflict and competition for resources, having a safe place to meet may be one of the most important preconditions for a participatory democratic process. One example, recounted during the New England workshop, illustrates the force of dominant notions of status in a hierarchical society and the retaliatory practices used to enforce the status quo. The workshop participant described the reluctance of forest contractors to display signs of wealth by, for example, parking a fishing boat in their driveway out of the concern that forestland owners with whom they contract will see the boat, conclude that they are making "too much" money, and reduce the amount they pay for contracted work. Another, perhaps more direct example of this process concerns forest workers whose immigration status may be uncertain or undocumented. These people, whose numbers are increasing across the United States and who in some regions already constitute the majority of forest workers, especially for labor-intensive aspects of forest management, are obviously reluctant to participate in any kind of organized public forum or meeting, yet their concerns are germane to community forestry, and the working conditions they experience are among the worst in the forestry sector.[12]

Language barriers also present challenges for full participation in many contexts. In this multicultural, multilingual society, inattention to the importance of translating documents into the languages of the various stakeholder groups, the importance of working with bilingual and trilingual people, and the need to provide simultaneous translation at public meetings invariably leads to the exclusion of those with limited English language skills. The Jefferson Center is among the few organizations involved in community forestry that has devoted the resources and energy to provide simultaneous translation services for the diverse gatherings it has sponsored.

Promoting participation of underrepresented groups in community forestry entails supporting communication capacities, networking, and organizational and leadership capacity development within and between groups. Examples include the provision of a safe or neutral place for groups to meet,[13] support for people to travel to relevant workshops, conferences, and meetings with elected representatives, and other ways of fostering communication and networking in underrepresented communities. An example of this latter point is the Pacific West Community Forestry Center–sponsored Hispanic worker conversations and meetings designed to facilitate communication, the exchange of forestry-related information, and the development of networking and participatory research capacities in the Latino community.[14] The importance of understanding and increasing the flow of information within and between communities was raised several times at the Pacific West workshop. It was suggested that there are multiple ways to facilitate intracommunity communication, including newsletters, radio, and community organization gatherings. Newsletters, group workshops, and trainings strengthen participant solidarity and community identity. Regular meetings facilitate communication and relay information. Communication helps build a shared understanding of challenges, barriers, and opportunities between groups. This, in turn, helps to increase groups' ability to act collectively to pursue shared interests. The most effective mode of communication varies between groups. These and other examples of mechanisms for facilitating stakeholder participation from underrepresented groups and communities illustrate some of the special considerations that are necessary for these groups to participate in the democratic processes. Rather than "equal treatment" under the rule of law, which translates into a system of advantage based on race, class, and other differences, special treatment of disadvantaged groups is necessary to promote full and participatory democracy.

Conclusions

The notion of community that emerges from this discussion is one that acknowledges the validity of the different perspectives and interests of diverse, empowered social groups and stakeholders. It is not a version of community founded on the myth of homogeneity or the principle of local autonomy but rather is based on common concern for ecosystem health and worker and community well-being.[15] Empowerment is central to this notion of community. Empowerment is a more participatory and democratic objective than autonomy. It is an open-ended concept that implies the ability to participate effectively in decision making. It is linked to justice, and thereby to democracy, through the development of institutional mechanisms that enable diverse people to participate effectively in decisions that affect them or the conditions under which they act. Empowerment "means, at a minimum, expanding the range of decisions that are made through democratic process," with "agents who are empowered with a voice to discuss ends and means of collective life, and who have institutionalized means of participating in those decisions, whether directly or through representatives, open together onto a set of publics where none has autonomy" (Young 1990:251).

This notion of empowerment avoids the exclusion and repression of difference associated with localism and autonomy while embracing the multilevel, fluid dynamic of effective participation within the diverse, multicultural setting of community forestry. It fosters engagement with the policy process at local, regional, and national levels and seeks ways to ensure that information, ideas, and representation flow as freely as possible between these levels. While acknowledging the importance and relevance of local knowledge and history and place-based community participation, it also avoids the pitfall of overemphasizing the local, thereby becoming vulnerable to nonlocal processes; in the larger economic context neither capital nor corporations profess allegiance to locality, capital flight and the corporate "race to the bottom" are characteristic elements of globalization, and seasonal movements of workers across space and sectors of the economy are a common feature of the social and economic landscape.

The idea of empowerment, in combination with the politics of difference, incorporates play and the enjoyment and excitement associated with interacting with others who are different. Young (1990:241) characterizes this element of the politics of difference in terms of a public that is "heterogeneous, plural, and playful, a place where people witness and appreciate diverse cultural expressions." This characterization of the politics of difference resonates with descriptions of truly collaborative and participatory community forestry processes. One well-known case is Newton County, Arkansas, in which seemingly entrenched conflict between property rights activists, back-to-the-landers, and other stakeholders was transformed through a series of workshops and meetings into productive and working collaborative relationships. Don Voth, a rural sociologist at the University of Arkansas and a key participant in the Newton County process, described the workshops and the overall process of shifting from entrenched conflict and distrust to a politics that acknowledged difference as legitimate. He said that what happened in the workshops was "weird and scary," that people were on the phone crying because of the emotional turmoil involved in shifting from a politics of exclusion to a politics of difference. Voth stated,

> The workshop was successful because a few people were willing to try something scary . . . the county extension agent played a pivotal role, really got carried away by the process and got into the fun of it all. The workshop demonstrated that people can stop fighting and get serious about being human beings. It demonstrated the importance of conflict resolution skills for both extension and Forest Service staff.

The emphasis here on being human, on fun, and on engaging with people who are different, whose very difference constitutes the elements of "weird and scary," clearly resonates with Young's discussion of the ways in which the politics of difference and the forms of participation by diverse groups it engenders lead to a public that is "heterogeneous, plural, and playful." Forest Service employee Crockett Dumas also described the play that results from opening up the process of community-based forest management and making it truly participatory for a heterogeneous group of empowered stakeholders. When he was district ranger on the conflict-ridden and occasionally violent Camino Real District of the Carson National Forest, an approach to bottom-up forest planning and management was developed that "got buy in from Native Americans, Hispanos, artists, hippies, trust funders, movie stars, strong environmentalists, and Forest Service employees."[16] The effort was so distinctive that the John F. Kennedy School of Government recognized it by giving the participants its "Innovation in American Government" award. In a striking echo of Young's discussion of play and the politics of difference (including an allusion to the dangers of excluding empowered stakeholder groups), Dumas commented, "You can either manage and have fun or else the alternative is the Watts riots."

CHAPTER 7

The Politics of Community Forestry

> The key is to place government in pursuit of renewal. . . . It means fundamentally rearranging the structural relations among market, state, and democracy so as to protect the environment and, at the same time, nurture democratic community.
>
> —*Williams and Matheny (1995:193)*

While community forestry practitioners, advocates, and supporters are innovating new forms of Jeffersonian political practices at the local and regional level, at the national level segments of the movement are effectively using the traditional mechanisms of representative pluralist democracy to advance the movement's goals. In Washington, D.C., members of national and regional community forestry organizations work with members of Congress and congressional staff on policy. Practitioners and supporters of community forestry also cultivate collaborative relationships with key federal agencies, and they speak with national interest groups. Long-standing national nongovernment organizations such as American Forests and the Pinchot Institute for Conservation also play a variety of supporting roles for community forestry within national legislative and policy-making arenas. Partly as a result of these efforts, the Forest Service is experimenting with collaborative models of resource management, some members of Congress and their staff members are receptive to and supportive of community forestry, and segments of the environmental movement are engaging with community forestry. As community forestry groups continue to work in the national policy arena, using the tools of interest group pluralism while straining to not become an interest group, the challenges of avoiding the exclusionary tendencies associated with those tools become all the more daunting.

In this chapter we address some of the national-level manifestations of the community forestry movement and review the ways federal resource management agencies, Congress, and national interest groups are responding to the movement's concerns. We examine the movement's implications for the organization and operation of the Forest Service, its ability to influence Congress and the national policy-making process, and its relationships with other national interest groups. We consider the tensions that arise when the drive for local empowerment hits centralized and hierarchical government structures and political processes. We discuss the critiques that some environmentalists have made of community forestry and identify possible areas of common interest that could constitute a basis for alliance building between community forestry practitioners and their organizations and environmental groups. Finally, we address, albeit in a cursory fashion, the relationship between the forest products industry and community forestry.

The Forest Service[1]

The dynamics of change within the Forest Service with regard to community forestry are complex and incremental. This is partly a reflection of the agency's institutional history, which embodies the Progressive ideal of hierarchically organized, expert-driven natural resource management. The process of change is also complicated by the fact that the Forest Service is internally heterogeneous. Therefore, as priorities in the agency gradually shift to become more supportive of community forestry, opposition by those who are skeptical of community forestry or who stand to lose resources and status as community forestry becomes more institutionalized will harden while the organizational positions of individuals and programs supportive of community forestry are strengthened.

As this section demonstrates, several elements are needed for the Forest Service to institutionalize its commitment to work collaboratively with communities in supporting community forestry. These include a clear, sustained commitment from Forest Service leadership in support of collaboration and engagement with community and worker issues. Absent the articulation of such a vision and commitment from the Forest Service leadership, and perhaps its further legitimization through legislation that renders Forest Service support for community forestry consistent with its other legal mandates, the momentum necessary for embracing community forestry ideas and objectives within the agency simply will not develop. Other necessary elements include the development of institutional mechanisms that build accountability into all levels of the organization. Unless engagement with community forestry is prioritized through an accountability and incentive system that is linked to process and results and is backed by strong leadership at the national, regional, and forest level, few Forest Service officers will choose to invest scarce resources in the challenging and time-consuming work of community forestry.

Analyses are needed of the specific types of organizational changes collaboration and community forestry entail. Based on these analyses, the agency's commitment to providing the resources necessary for the changes and rewarding those who embrace them can be clearly communicated to all agency employees. Thus, for example, preparing, administering, and monitoring a large number of small contracts (for which smaller, more community responsive contractors can more successfully compete) take significantly more staff time than a few large contracts and often cost more on a per acre basis. Unless staff members are supported for developing more innovative and time-consuming contracting mechanisms and for choosing other criteria in addition to cost-effectiveness to evaluate bids, the disincentives associated with innovation will prevent most contracting officers from stepping outside the box. Part of the process of mobilizing the necessary political and agency support and resources for undertaking these changes requires the development of compelling arguments for why community forestry should be supported. Articulating the ecological and social benefits of collaboration and community forestry and practicing community forestry on public lands through a series of pilot projects that are monitored through all-party or multiparty monitoring frameworks will go a long way toward building the political momentum and agency support necessary for the Forest Service to embrace community forestry in a systematic manner.

Until the systemic changes described here are achieved, community forestry in the Forest Service will remain the domain of risk takers, will persist at the level of pilot projects, and, though perhaps supported by some segments of the agency, will not be adopted as a central part of its mandate by the agency as a whole. Despite the significant challenges that lie before it, the Forest Service has initiated institutional change to support community forestry in several ways. Examples include increased attention to the full array of Forest Service contracting authorities and the development of new contracting mechanisms such as "best-value" contracting, implementation of the national pilot stewardship program on 28 sites around the country, the development of a collaborative support team based in the Washington Office of the Forest Service, and the creation of draft regulations that mandate and support community collaboration. Most recently, as a result of the passage in October 2000 of the Secure Rural Schools and Community Self-Determination Act (PL 106-393, known as the "County Payments Bill"), the agency has become a key player in newly established county-level Resource Advisory Committees that, based on the recognized interdependence between forests and rural communities, evaluate and disburse funds for projects oriented toward restoring forests and enhancing community well-being. In addition to these forms of institutional change, some Forest Service employees operating within the existing organizational structure of the agency, and perhaps pushing the envelope of that structure, have developed innovative ways to collaborate with local communities.

Cooperative Forestry

There are many diverse points of engagement with community forestry within the Forest Service. The two primary arenas of engagement are the National Forest System and the Cooperative Forestry branch of State and Private Forestry. Whereas the former is charged with managing the national forests, the latter has a diverse portfolio that includes Economic Action Programs, Landowners Assistance Programs, and the Urban and Community Forestry Program. These programs, as well as other Cooperative Forestry programs, are managed in part through partnerships with state forestry organizations, local governments, and universities. The objectives of programs such as the Rural Community Assistance Program (one of three components of the Economic Action Program) and the Urban and Community Forestry Program directly support those of community forestry. For example, Rural Community Assistance projects, which tend to be concentrated in rural areas, often are implemented in a decentralized manner in close cooperation with local communities. Most Rural Community Assistance projects include community capacity-building goals. The biannual Rural Community Assistance conference highlights the accomplishments of rural communities made through their partnership with Cooperative Forestry and participation in Cooperative Forestry programs. The conference also helps communities network with and learn from each other and access available funds, expertise, and other resources.[2]

The Urban and Community Forestry Program, authorized in 1972, funded in 1978, and accorded a high priority in the 1990 Farm Bill, enables the Forest Service to work collaboratively with state forestry associations, nonprofit organizations, and other professional and conservation organizations to enhance and restore urban ecosystems. Focusing on urban areas, much of this work is carried out with a strong emphasis on community participation and empowerment. In some cases projects funded by the Urban and Community Forestry Program include a focus on brownfield development, economic revitalization, and environmental justice.[3]

Cooperative Forestry has a long track record of collaborating and promoting collaboration with communities in ways that empower communities and individuals and build community capacity. However, there are significant limitations on Cooperative Forestry's ability to serve underrepresented and underserved communities, including landowners, forest users, and workers, and on its ability to promote an agencywide commitment to working collaboratively with communities, especially minority communities. The first limitation has an important opportunity embedded within it. It concerns the agency's capacity to respond to the needs of minority forestland owners, users, and workers and to address workforce diversity and civil rights issues within the agency. These issues are inextricably linked. As the Forest Service becomes more responsive to internal multicultural issues and thereby develops into a more multicultural organization in which civil rights issues of justice and equity are valued and upheld, it will be better able to meaningfully engage with the diverse communities and people of color that have forest claims. The Forest Service, well known for its slowness in achieving workforce diversification, launched a series of initiatives in the late 1980s and early 1990s to help it become more multicultural and better able to address social justice and equity issues. These included organizing a National Diversity Conference in 1990, appointing a Taskforce on Workforce Diversity, introducing the *Toward a Multicultural Organization* report in 1991 (which identified 11 workforce diversification goals for the agency at both national and regional levels), and, in 1992, preparing the National Forest Implementation Plan for the *Toward a Multicultural Organization* report. Efforts to implement and institutionalize the recommendations of these and other reports at all levels of the agency have dwindled during the intervening years. The reasons are varied. They include budget constraints, agency downsizing, preoccupation with ecosystem management and other resource management issues, opposition to affirmative action within the agency, the absence of a system for holding employees accountable and rewarding them for their multicultural achievements, and a lack of prioritization of these issues from top agency leaders (Silva et al. 1988:8–12).

To the extent that progress has been made with respect to issues of workforce diversity and multiculturalism within the agency, it has resulted largely from the efforts of the Forest Service's civil rights unit and those Forest Service employees for whom civil rights issues are a high priority. The efforts of both the unit and other committed Forest Service employees are based on the assumption that only a diverse and multicultural agency can properly serve the interests of all people in the United States. To facilitate accomplishment of these twin goals, a variety of special emphasis programs have been developed. These include the African American Program, Native American Program, Asian Pacific American Program, Disability Employment Program, Federal Women's Program, and Hispanic Emphasis Program.

Additionally, in 1998 Memorandums of Understanding were developed between the Forest Service and organizations such as the National Organization of Black County Officials, National Black Farmer's Association, and the U.S. Department of Agriculture (USDA) Coalition of Minority Employees. These Memorandums of Understanding establish broad areas of mutual interest, outline the responsibilities of signatories to each other, and identify how they can advance each other's respective missions. Members of the civil rights unit work to raise awareness of multicultural issues within the Forest Service, bridge between minority communities and the agency, and diversify the Forest Service by engaging in outreach to minority youth in schools. Despite the diligent and determined efforts of the civil rights unit and other Forest Service employees, few would disagree the Forest Service has a long way to go before it achieves the goals laid out in the *Toward a Multicultural Organization* report. Until it does, divisions such as Cooperative Forestry will face an uphill struggle in their attempts to work with and advocate wider agency engagement with rural communities, especially communities of color.

The second constraint stems from the fact that in most cases Cooperative Forestry does not directly develop and implement community-based projects. Instead, it funnels funds and ideas through partner organizations, particularly state forestry agencies and land grant universities. State forestry organizations often are embedded in long-standing relations that tie them to specific programmatic objectives and associated client groups. Whether the focus is on game animals as a primary component of wildlife management or extension and outreach to large industrial forestland owners, state forestry programs must struggle to redirect their energies toward ecosystem management and to embrace the full array of forestland owners, users, and workers.[4] Without strong grassroots and community support and state-level political support, there are few viable opportunities for state foresters to redirect their programs to be more consistent with the objectives of community forestry. This has been identified as a particular challenge in the 13-state Southeast region, an area in which the lack of targeted outreach to African American landowners (agricultural and forest) has fueled the long-standing distrust of the USDA among African Americans, hastened African American land loss in the South, and resulted in a successful class action suit against the USDA by African American landowners. Even when state foresters would like to implement more progressive forestry outreach and extension programs, they may not be able to because they lack the necessary political space and freedom to chart a new course in terms of program objectives. Because of such institutional roadblocks at the state level, control of some Cooperative Forestry funding has been retained by the Forest Service and directed toward community capacity-building endeavors, which sometimes are less project driven. In some cases, the move to retain control over funds previously allocated to state forestry organizations has predictably led to conflict between Cooperative Forestry and individual states.[5]

A third limitation on Cooperative Forestry's ability to reach rural communities and promote community forestry stems from the lack of articulation between programs such as Rural Community Assistance and the activities and programs of the National Forest System. The disconnect between the National Forest System and Cooperative Forestry results from several factors. One is the historical organizational culture of the Forest Service, in which the National Forest System and national forest management were considered the bread and butter of the agency's mandate and purpose whereas other divisions, such as State and Private Forestry, were less central. This disconnect can translate into a lack of field coordination between Cooperative Forestry and National Forest System programs. One rather extreme example of this was a situation in which a district ranger tried unsuccessfully to encourage community involvement in the forest planning process, while in the same community and unbeknownst to the ranger, there was a thriving Rural Community Assistance program that had developed excellent working relations with the community leaders. The absence of coordination between the Rural Community Assistance program and the work associated with the National Forest System precluded a productive and mutually beneficial partnership and also thwarted the district ranger's community outreach attempts. In the Pacific Northwest, some community economic development and worker retraining projects that were part of the Northwest Forest Plan struggled or failed outright because they depended on a link to national forest system management that did not exist (Kusel et al. 2002). Rural sociologists who have studied the Rural Community Assistance program argue that the relationship between state and private forestry and the National Forest System must be restructured (Frentz et al. 1999). Rather than the various (non–fire-related) programs of state and private forestry, especially those of Cooperative Forestry, being perceived as secondary to the core mission of the Forest Service and therefore vulnerable to budget cuts (as evidenced in the Bush administration's proposed 2003 budget, which eliminates the Economic Action Program), community assistance, planning, economic development, and related programs of Cooperative Forestry must be fully integrated with the mission of the National Forest System. This would help reconnect the communities and workers to the forest and make community involvement in forest planning a major responsibility of the system (not relegated to a function of Cooperative Forestry), and it would tighten the linkages between forest management and rural community economic development and well-being. It would entail building positive relations with rural communities and workers adjacent to national forests before initiating public review of forest plans, management, and planning processes. Public involvement structured along these lines would lead to more deliberative discussions and result in an increased public engagement with and stake in the forest plan. Instead of the district ranger having to defend the plan, this kind of engagement could lead to a willingness on the part of local communities to defend "their" plan.[6]

The National Forest System: How to Institutionalize the "Radical Center"

The National Forest System is the second primary arena of engagement between community forestry and the Forest Service. The structure and content of that engagement focuses on public involvement in forest planning and contracting for a wide variety of services ranging from carrying out silvicultural prescriptions to fuels reduction and campground management. Community forestry is intimately associated with both of these aspects of national forest management. The first set of issues—those concerning forest planning—relate centrally to the challenge of how to resolve the often intense conflict surrounding different, and at times competing, visions of how and why a national forest should be managed. The second set of issues—those concerning the organization of work on the national forests (including the harvesting and processing of nontimber forest products)—are of central importance to communities whose livelihood depends on work in the woods.

Community Involvement in Forest Planning

Some of the best-known stories of community forestry on public lands concern innovation with regard to local community involvement in forest planning and management. Catron County, southwest New Mexico; Newton County, Arkansas; the Ponderosa Pine Partnership in southwest Colorado; and the Carson National Forest in northern New Mexico are just a few of the places that have become well known for the innovative and creative conflict resolution processes that were developed to move beyond debilitating and at times violent confrontations between groups that held opposing visions of how the national forest in their area should be managed. In many cases the way out of gridlock embodied principles of face-to-face public deliberation and Jeffersonian democracy.

The stories of local-level innovation with regard to national forest planning processes, and the ways in which both stakeholders and Forest Service personnel were willing to take risks to break the gridlock and reintroduce civility into what had in many cases degenerated into tense and violence-prone situations, are marvelous in their own right and deserve to be chronicled; indeed, many have been described.[7] Our primary purpose in reviewing two examples of innovative forms of agency-facilitated public participation in public lands management is to illustrate some of the new ways government can function at the community level and to suggest that the kinds of government innovation that have occurred in these situations might be more broadly institutionalized within the Forest Service and other public land management agencies such as the Bureau of Land Management (BLM). Recognizing that we have only partial views of these processes because of the limited number of participants interviewed and wary of the risk of extrapolating from isolated examples general trends and patterns of change that do not yet exist, the following paragraphs present mostly an agency perspective on community collaboration.

Some federal public lands managers and their community partners are experimenting with what might be called a radical new approach to working with the public on public land management issues. In many cases their experimentation was driven by the frustration associated with gridlock, socially destructive levels of conflict within rural communities, and the "whipsaw" effect of shifting political winds and priorities. Often, breaking up the gridlock entailed a radical revisioning of the role of government in resource management planning processes. It involved devising ways to work with multiple stakeholders that gave the public more ownership in the outcomes of the planning process and helped develop and strengthen the land stewardship ethic in rural communities. This entailed risk taking on the part of field staff and relinquishing tight control of the planning process. As described by Crockett Dumas, former district ranger in the Camino Real District of the Carson National Forest in northern New Mexico, it involves halting the practice of "hiding behind the green badge," being willing to "step through the line of control" to empower people to assume joint responsibility for ecological stewardship, and "learning to find the appropriate way to say 'yes.'"

These approaches, born of attempts to transcend intractable conflict and gridlock and based on community-driven planning processes, stand in stark contrast to the formal models of public participation that generally govern the structure of public involvement in public lands planning processes. The combination of intense pressure on line officers to get forest plans approved and the restrictions of formal participation procedures prevented meaningful participation.[8] Dumas noted that the process "stole the dignity" of the public as line officers basically bargained with different sectors of the public, asking them to give something up on one issue to get something on another. Noting the limitations of this approach, he said, "When you force people to compromise, the results are rarely lasting."

Within the parameters established by relevant legislation and policy, innovators such as Dumas advocate sharing power, decision-making authority, and information with the various stakeholders and communities that have interests in a particular national forest or BLM district. Rather than proposing an array of already developed planning options and soliciting feedback on these agency-derived plans, they advocate an approach in which people develop their own preferred planning, strategy, and management options. This participatory approach to public lands planning and management gets the public involved during the initial stages of the planning process rather than only at the back end. To paraphrase Dumas, the basic objective is to get the public to tell the Forest Service what it thinks should be done; this approach of following rather than leading requires line officers to develop listening skills: "First and foremost, you have to listen—you can't argue until you first have developed rapport." This entails providing the decision-making space for people to express what they want to see done on the forest and then finding ways to move in the directions so defined. When it works, management priorities determined through grassroots, community-based processes float up through the public lands council and county commissioners to state- and national-level leaders and representatives. Other innovators have noted that these approaches involve developing a "deeper" democracy, a "learning model of democracy" in which citizens are empowered to be part of government and scientists are not just educators but also begin to learn what are community needs and then develop research projects and agendas that respond to those needs.

The results of these approaches are impressive. Dumas and his colleagues and partners collaboratively developed a forest plan that got buy-in from a very diverse spectrum of stakeholders and Forest Service employees. As Dumas noted, "The management of natural resources is totally due to whom you engage with, the public and employees . . . how they get ownership, share power and decision making." In recognition of their efforts to develop a community-based strategy to implement ecosystem management on the district, the Carson National Forest Core Team and the Camino Real Ranger District were awarded Vicc President Gore's Hammer Award in 1997. This award is designed to recognize teams of federal employees (and those with whom they collaborate) whose work results in a government that "works better and costs less." In 1998 they won an Innovation in American Government Award from the J. F. Kennedy School of Government at Harvard University. A video recording of what occurred on the Camino Real Ranger District, "Good for the Land, Good for the People," has been made to document and disseminate this community-based, collaborative approach to natural resource planning. Both Dumas and another innovator, Gary McVicker, ecosystem manager for the BLM in Colorado, have developed short training seminars that they take on the road in an attempt to share with other federal employees and communities their insights concerning community-based resource management. Dumas (until recently) gave two seminars every month on how what has come to be called "collaborative stewardship" evolved and how it was applied on the Camino Real Ranger District. In a similar vein, McVicker, along with two other (non-BLM) colleagues, has put together a 3-day workshop on collaborative stewardship, which has been held in more than 20 communities over the last 3 years. The purpose of the workshop is to convey an understanding of the new relations that are possible between rural communities and government vis-à-vis resource management on public lands, to rebuild trust between rural communities and government, and to plant the seeds at the grassroots level that will eventually generate the political pressure on agencies and policy makers in Washington, D.C., to support the bureaucratic and policy reforms necessary to fully achieve the vision of shared, collaborative decision making these innovators are advancing. McVicker has also worked with a colleague at the national BLM training center in Phoenix, Arizona, to design training programs in collaborative stewardship for BLM employees.[9] These seminars, workshops, and training programs all represent efforts to institutionalize community forestry on public lands.

As argued at the beginning of this chapter, institutional change of the kind needed to support community forestry is a work in progress. Certainly the idea of collaboration is much more generally accepted within the Forest Service and BLM than even just a few years ago. Forest Service planning regulations, public statements by agency leaders that support collaborative working partnerships with rural communities, the higher profile of Cooperative Forestry programs, Forest Service white papers and reports promoting collaboration, and the stewardship pilot project program are all evidence that change is occurring. However, systematic institutional change of the kind that would provide full organizational encouragement and support for line officers to engage in innovative collaborative conservation initiatives has yet to take place.[10] Although the track record of successful innovation is clearly visible, the specific tools of those innovations have yet to be institutionalized. Nor has the organizational reward system of the Forest Service been tailored to reflect and value the new skills and abilities that successful collaboration entails. On the contrary, the proposed elimination of the Economic Action Programs of state and private forestry and the recent turn towards a "business model" of organizational structure in which services are provided to "clients" on a fee basis run counter to the notion that people are part of government and should therefore be part of a collaborative decision-making process based on principles of partnership and adaptive learning. In short, the deeper structural changes that are needed to support community forestry have not yet occurred, and it is unclear whether the political will exists to do so.[11]

The Organization of Work in the Woods

Who works in the woods and the conditions under which that work is carried out is a second important component of community forestry on public lands that complements the focus on community-based forest management and planning. The organization of work in the (public) woods crystallizes many issues concerning who benefits from community forestry and who does not. Workforce differentiation, the variety of forest work, the institutionalized patterns of contracting for forest work, and the variety of alternative and newly developed contracting mechanisms contribute to the complexity of the issue. Key debates within community forestry concern both working conditions and compensation and issues relating to which groups of workers are able to access forest work. These debates occur within a broader context of downward pressure on wage rates, labor laws that are difficult to enforce, and the pervasive view that occupation is a weak basis for forest enfranchisement. These conditions make it particularly difficult to build bridges between different groups of forest workers, further challenging attempts to build capacity among forest workers and articulate a common vision of the needs of forest workers.

The people who work in the woods are a highly differentiated group. In addition to their racial, ethnic, and linguistic diversity, forest workers are also highly differentiated in terms of how long they or their families have been involved with forest work and their legal status (as citizens or documented or undocumented immigrants). The precarious legal standing and weak rights of undocumented workers render them particularly vulnerable to exploitive employment conditions and compensation. Forest work itself is also highly differentiated. It ranges from projects that involve heavy equipment but not many people to tree planting, thinning, and other tasks that are labor intensive. Contracting procedures on public lands historically have been structured to minimize costs and maximize on-the-ground results per unit investment. Therefore, forest work often has been packaged in large, easy-to-administer units and contracts awarded to the lowest bidder. Contracts, especially for labor-intensive tasks, are awarded to wide-ranging contractors who move crews of workers long distances from contract to contract. Analysis of national forest contracting in the Pacific Northwest indicates that contractors that mobilize crews for labor-intensive forest management tasks or that do highly skilled work (e.g., helicopter logging) travel farther than contractors who bid on jobs that use heavy equipment (Moseley and Shamkle 2001:37). It is particularly difficult to monitor the working conditions and ensure compliance with labor laws governing working conditions and compensation when forest workers are part of mobile crews that cover large areas and never stay long in one location.

The issue of hiring local versus nonlocal forest workers and which groups of workers are getting the work often is debated within community forestry. Many supporters argue that to support economically depressed rural economies, contracts should be repackaged in sizes and ways that enable local contractors to successfully bid on them. Achieving this entails a number of institutional changes in public agencies' contracting processes and criteria for evaluating contract bids. It also requires the presence of a stable and skilled local workforce willing and able to do the variety of tasks associated with forest management. In rural areas where the amount of forest work has declined steeply because of reduced timber harvest levels, the presence or availability of a local workforce cannot be taken for granted; many forest workers have outmigrated in search of jobs or changed fields altogether.

To date, there have been few attempts to provide coherent policy guidance to line officers, especially contracting officers, regarding how to navigate through the challenging issues and questions that these issues present. On the other hand, diverse and innovative programs attempt to address some of the challenges associated with workforce issues. These include the 28 congressionally authorized pilot stewardship projects, the introduction of "best-value" contracting procedures, experimentation with trading goods for services on forest restoration projects, the Jobs in the Woods program, and other ecosystem workforce training and employment programs.[12] However, most of these programs and procedures remain at the level of experimental innovation or do not have ongoing funding support. Unless a broad agency mandate exists to support their institutionalization, they remain vulnerable to being marginalized by those who are not convinced of their merit. Achieving these goals will entail enabling state and federal legislation, administrative support within public agencies that combines accountability mechanisms with incentives for these forms of contracting, willingness to allocate scarce agency resources to working with more small-scale contractors, and determining explicit guidelines for evaluating different "best-value" bids.

From an equity and social justice standpoint, the diversification of contracting programs and procedures to respond to concerns about promoting community-based economic development in rural areas must be accompanied by a broad-based analysis of the full spectrum of forest labor issues. Such an analysis is needed to ensure that the gains of one group of forest workers do not come at the expense of other, less enfranchised workers. Mobilizing the support necessary for developing the policy and program instruments needed to protect all forest workers from exploitive labor relations and fostering the capacity among forest workers to resist exploitation are some of the biggest challenges facing the community forestry movement. Because of the differentiated nature of the workforce, the diversity of tasks associated with forest management, and the wide variety of contracting mechanisms, it is likely that different ways of organizing work in the woods will persist into the foreseeable future; therefore, although the proportions may vary, both local and nonlocal forest workers will continue to be involved with various aspects of forest management. This highlights the importance of strengthening the enfranchisement of all forest workers and developing the policy and program tools necessary to make that enfranchisement meaningful. Some of the common goals of these programs include providing continuous employment opportunities for forest workers with remuneration at family wage levels, packaging work contracts in sizes that enable small-, medium-, and large-scale contractors to bid successfully, developing innovative service and timber sale contracts that promote economically viable utilization of small-diameter materials, making explicit links between models of forest work and product utilization and the growth of viable small rural business enterprises, and developing a trained and certified ecosystem restoration workforce that has access to long-term employment opportunities. Although these goals are integral to achieving the broad objectives of community forestry, it is clear that significant progress at a variety of levels remains to be made.[13]

The Extension Service

It is hard to underestimate the importance of the role of the extension forester in community forestry on private forestlands. Extension foresters, whether county foresters associated with the state forestry bureaucracy or part of the land grant college and university system, mediate between forestland owners and government programs that promote forest stewardship and conservation, helping to communicate landowner concerns, constraints, and needs to relevant policy, administrative, and research arenas and occasionally working as a catalyst to foster collaborative networks of forestland owners. In this respect extension foresters can play an important catalytic role in forwarding community forestry on private lands. On the other hand, extension forestry can be embedded in historical relations that link its roles and program priorities to the interests and needs of the forest industry or to other program goals that may not be consistent with community forestry. In this respect the track record of extension forestry is somewhat analogous to that of the Forest Service: Although innovative and risk-taking extension foresters have effectively supported community forestry initiatives in their own regions, the systemic changes necessary to institutionalize those initiatives and provide broad organizational support for community forestry have yet to evolve.

Predictably, the regions where extension forestry is most central to community forestry are those with high proportions of privately owned forestland. Forestlands in the Deep South, Appalachia, Northeast, and Midwest and Lake States, though possessing important areas of publicly owned forestland, are predominantly privately owned. These areas contain many examples of the ways in which extension forestry can promote community forestry. Extension and county foresters develop and strengthen positive change by conducting educational workshops and training programs and helping to catalyze networks and associations of nonindustrial private forestland owners. Extension and county foresters interested in advancing community forestry promote mutually beneficial relationships between nonindustrial forestland owners and local, secondary wood products manufacturers, support independent third-party forest certification, and encourage the adoption of regional best management practices.

Despite these gains, the traditional orientation of parts of the forestry extension service is inconsistent with the philosophy, objectives, and methods of community forestry. In some cases, for example, extension forestry is aligned with the research and extension needs of larger, industrially oriented forestland owners, thus overlooking the needs and management priorities of minority and resource-poor nonindustrial landowners. This can create a lack of symmetry between the needs of nonindustrial landowners and the activities of extension foresters. Research conducted by Auburn University (Alabama) rural sociologist Conner Bailey, interviewed as part of this project, identified the discrepancy between the forestry extension priorities of the Public Advisory Councils and the actual activities and programs of extension foresters. This research was used to help reorient the extension service's priorities.[14] Identifying the gap between the needs of rural communities and the activities of extension agents is one step toward bringing the latter more in line with the former. On a broader regional level, rural African American landowners, agriculturalists, and forestland owners alike are commonly distrustful of the federal government and the USDA in particular. In many areas there is very little interaction between the activities and priorities of extension foresters and African American landowners. In these areas attempting to reestablish trust and communication with minority landowners is a first step toward achieving community forestry goals.

Nonprofit community forestry organizations can facilitate the process of building communication, trust, and working relations between minority and underserved landowners and extension forestry organizations. In the southern United States the Federation of Southern Cooperatives and its affiliate organizations have been working on these issues. Minority and resource-limited landowners are acutely aware of the inequities in the provision and availability of government programs, loans, and subsidies. For example, in Mississippi, which has a minority population of 43 percent, minorities own 28 percent of the land base, yet they receive only 6 percent of the government programs for forestry (Taylor 2001). The primary focus of the Minority and Limited Resource Landowner Forest Program at the Federation of Southern Cooperatives is to increase the access of these landowner groups to government programs that support forestry and agriculture throughout the South. Many landowners, having suffered financially from the volatile markets for agricultural products in recent years, are turning to forestry to complement agriculture. In some cases, landowners have formed local cooperatives to advance their collective interests. For example, the Winston County Self Help Cooperative, founded in 1985 and a member of the Mississippi Association of Cooperatives, which operates under the umbrella of the Federation of Southern Cooperatives, was formed "for the purpose of helping to serve limited resource farmers with technical assistance and management and to build unity among small farmers" (Taylor 2001). Many of the 17 families that make up the self-help cooperative have diversified into timber production and management. Some of the cooperative's goals include increasing the number of underserved and limited-resource family farmers that participate in USDA and state programs, improving the participation of these landowners in the county committee election process, increasing cooperation and developing better working relations with other state offices, lobbying for more state support for nonindustrial private forestry, and supporting the production, harvest, and processing of nontimber forest products by minority landowners (Taylor 2001:1). The bottom-up efforts of these grassroots-based organizations to build local capacity and demand the government services minority and limited-resource landowners deserve are complemented by growing political pressure from above to address the historical and institutional legacies of biased outreach and program availability.[15]

Despite the achievements of organizations such as the Federation of Southern Cooperatives and the Winston County Self Help Cooperative, significant barriers remain. Progress in Alabama is uneven across the state and is certainly not matched with equal progress in other states in the South. A lack of estate planning (to help provide continuity of land ownership), challenges associated with absentee ownership (that result in below-market timber and land sales to unscrupulous buyers), unmet research needs (e.g., the economics of woodlots), the need to involve younger generations in forestry and land management, and ongoing distrust of government are still characteristic of the conditions of minority and resource-poor forest management in the South.

Community Forestry Politics at the National Level

At the level of national politics, the community forestry movement has the potential to help forge a bipartisan approach to natural resource management issues. For example, in many western states the Democratic party has been on the wane because of its association with, from the perspective of rural, resource-dependent communities, top-down preservation-oriented, antidemocratic policies and legislation. The Republican party, on the other hand, has been associated with championing the extractive interests in the West, regardless of the associated environmental impacts. In many parts of the rural West, it is the dominant party. Kemmis (2001:225) notes the ironic consequences of this for community-based collaboration in the West: "There is something paradoxical about the fact that in the West, the region of the Democratic Party's greatest weakness, there should be such a vital, growing democratic movement and that it should be overlooked by the party." Kemmis (2001:229) argues that if the Democratic party could hark back to its Jeffersonian roots in participatory democracy and if the Republican party could help westerners articulate their principles of environmentalism that developed "from living on and loving well the landscapes they are now prepared to steward," then the collaborative approach to natural resources management might create a space in which "western Democrats and Republicans could . . . harmonize their voices in support of a maturing western agenda."

Over the last 10 years practitioners and supporters of community forestry, working through their respective regional and national community forestry organizations, have begun to strengthen the viability of the bipartisan political space to which Kemmis refers. Using vehicles such as the ongoing series of meetings in Washington, known as the "Week in Washington," press releases, testimony at congressional hearings, and coalition building with environmental organizations and others, community forestry advocates have been able to effectively and forcefully articulate a view of forestry that transcends the polarized jobs versus the environment debate and forward a vision of forestry that includes both. This message has been met with an increasing degree of bipartisan support in Congress, especially from western states' congressional representatives. One of the net results of this process is that some congressional leaders and their staff have begun to advocate for community forestry. This advocacy is reflected in national legislation supportive of community forestry such as the National Fire Plan and the 2001 Department of Interior and Related Agencies Appropriations Bill (which funded the plan and mandated unprecedented levels of local involvement and collaboration as part of its implementation). Other outcomes of the increased visibility of community forestry in Washington have been the Forest Service's increased interest in community collaboration as well as community forestry's increasing acceptance among some national environmental organizations such as The Nature Conservancy and Environmental Defense, especially with respect to their growing interest in community-based conservation initiatives.

The increasing national stature of the community forestry movement, along with the circulation in Washington, D.C., of key community forestry ideas such as local collaboration, raises interesting issues and some potential pitfalls. One is the concern that although legislators may embrace the language of community involvement and local collaboration, what they mean by "involvement" and who, in their mind, constitutes the "community" may differ from the interpretations of community-based practitioners. This is particularly relevant because these are new (and sometimes ill-defined) ideas. The embrace of these terms by congressional staff and legislators represents a dramatic departure from business-as-usual for agencies and people accustomed to more centralized and hierarchical approaches to resource management. Although advocating increased public involvement and community collaboration seems consistent with community forestry, clearly defining what they mean, how they are to be achieved, and how they are to be funded are essential to the success of the endeavor. Simply calling for them without providing the means to achieve them, in the long run, will undermine their realization.

A second possible pitfall associated with developing a more high-profile presence in Washington is the possibility that vested and partisan interests may appropriate and use for their own purposes the message of community forestry, perhaps shorn of its equity, social justice, and environmental component. The potential for this increases as community forestry achieves greater visibility in Washington policy circles. As the political visibility and currency of community forestry increase, so does the temptation to co-opt its bipartisan message and use the language of community forestry for partisan purposes. The threat of co-optation is closely related to the concern that arises whenever a grassroots-based movement begins to make inroads into a national political arena dominated by interest group–based democratic pluralism. As some community forestry groups begin to develop ties with congressional leaders and their staff, special vigilance must be exercised to preserve the broadly inclusive and participatory nature of the movement. Only a strong self-conscious commitment to participatory democracy and social justice will provide the necessary fortitude to resist the temptation to sacrifice long-term equity for pragmatic short-term political gains and influence. As high-capacity community forestry organizations gain political access to the policy-making process, ensuring that the forms of community forestry policy that emerge from that process reflect ongoing commitments to equity and social justice becomes an increasingly important challenge.

National Environmental Groups and Community Forestry

Seeking common ground and the advancement of joint interest, community forestry practitioners and supporters have made and continue to make efforts to build relationships with national environmental organizations. These efforts have taken diverse forms. They have included conferences, workshops, field tours, and seminars as well as formal and informal meetings locally, regionally, and nationally. Community forestry practitioners have sought to communicate their objectives to environmental organizations through these forums in the hopes of identifying areas of overlapping interest, more effectively responding to the concerns of environmentalists, and building a broader base of support for the movement. Although the term "national environmental groups" covers a broad and diverse range of organizations, in general many of them continue to view the community forestry movement with some degree of skepticism, if not outright opposition.[16]

One of the primary sources of friction between the community forestry movement and many national environmental groups concerns efforts to increase local decision-making authority through collaborative institutional arrangements for resource management on public lands. Environmental organizations are reluctant to relinquish power to local forums and are concerned about the possible antidemocratic nature of local processes. Environmental organizations worry that many of the hard-won battles at the national level against overexploitation of public lands resources by commodity interests may be lost if decision-making authority is shared with rural communities and groups through community-based collaborative processes that, they fear, might weaken the enforcement of important national environmental legislation.

Some environmental organizations argue that local decision-making forums are antidemocratic or exclusionary because they tend to exclude marginal or peripheral local groups (as happened in the implementation of the Taylor Grazing Act), and they disenfranchise broader publics, communities of interest, and national organizations who also have rights in these public resources but who may be unable to participate in local-level collaborative forums (Blumberg and Knuffke 1998). They also question the long-term viability of voluntary, time-consuming decision-making processes, and they assert that such forums are incapable of addressing large-scale ecosystem or regionwide landscape issues.

Critics of community forestry worry that devolving planning authority and decision-making power to local arenas will enable corporate commodity interests to more easily access and extract resources from public lands.[17] They argue that rural unemployment, local dependence on the resource base for sustaining rural livelihoods, and inexperience in thwarting powerful commodity interests render local forums vulnerable to co-optation by corporate representatives seeking to maximize short-term profit at ecological and social expense. Indeed, the procommodity bias imputed to local communities has led some writers to argue that the efforts by community forestry practitioners to reduce gridlock in rural areas is actually aimed at "getting timber production back into high gear" (McCloskey 1999:626).

Coggins (1999) offers other related critiques of increased local community involvement in public land management. He asserts that past experiences with local control of public resources have resulted in short-term profit and overextraction. He argues that local control represents abdication by public agencies of responsibility for making decisions they are mandated to make. He also makes a sociological argument that local groups are insular, resist economic and political change, and seek a return to a mythic past by invoking terms such as "tradition," "lifestyle," and "culture and custom."

Environmental groups such as the Sierra Club and legal scholars such as Coggins use these critiques as the basis for their opposition to community forestry on public lands. They argue that the current political system (i.e., existing legislation and division of authority between executive, legislative, and judicial branches of government) works well "for the most part," and the existing policy-making process is inherently representative because "anyone . . . can have a say" (Coggins 1999:610). Devolving authority to local forums violates agreed-upon principles of pluralistic politics in which the interests of the majority prevail over the minority. Current deficiencies, where they exist, would go away if "Congress actually decides the political resource allocation questions; the executive carries out the letter and spirit of the law; and the courts make sure the executive does just that" (Coggins 1999:610). Accompanying this support for status quo policy-making processes and regulatory approaches are assertions that resource extraction as the basis for rural livelihoods is declining anyway and that local resource-dependent communities have not been negatively affected by those declines because of burgeoning tourism (Coggins 1999:28).

Why do opponents of community forestry reject community-based initiatives in favor of a strengthened political status quo model in which agency discretion would be reduced, congressional oversight increased, and implementation and enforcement of national environmental legislation strengthened? One possible explanation for the continued adherence to status quo policy processes and opposition to rural community-based resource management initiatives could be the relative success national environmental groups have experienced in status quo political arenas, particularly their recent ability to repel the explicitly antienvironment agenda of the Republican-dominated 104th Congress. This success has resulted from a combination of the support from the urban "checkbook" environmental community and public opinion polls that consistently show that protecting the environment is a widely shared value, which politicians disregard at their own risk. When a set of interests is able to develop and maintain dominant influence by marshaling resources such as votes and money for lobbying, they will strive to keep the locus of decision making in the arena where their resources are most effective and where they are most able to influence agenda setting and rule making (Gaventa 1980).

However, as noted earlier, the politics-as-usual model also has increasingly unacceptable shortcomings, many of which led to the emergence of community forestry in the first place. As discussed in Chapter 4, these include the way pluralism promotes acrimonious litigation, weakens discourses of civil public engagement, and generates gridlock. Furthermore, it does not produce lasting solutions because of the short-term vagaries of 4-year election-based cycles of political action. Centralized, hierarchical forms of bureaucratic organization, the monopoly on valid knowledge of "neutrally competent" scientific experts, and the reliance on top-down planning models are also ineffective in decision-making contexts characterized by increasing complexity and uncertainty, especially when stakeholders possess divergent value systems and management options are constrained by the legacies of historic resource degradation (Nelson 1996; Cortner and Moote 1999).[18]

Seen in this light, despite the opposition of some environmental organizations, collaborative community-based approaches to managing public lands resources can be viewed as important harbingers of the next era of public land management. Supporters of community forestry, including those from within the national environmental movement, argue that local involvement in public lands management is especially important under the increasingly prevalent conditions of complexity, uncertainty, divergent value systems, and the need to channel scarce funds toward ecological restoration. They see parallel shifts within the business world in which increasing information and flexibility needs for effectively interacting with an unpredictable and changing environment have wrought fundamental organizational changes; primary among them is the supplanting of hierarchical with network-based models that emphasize horizontal information flows and the devolution of decision-making authority. Some argue that community-based resource management initiatives are crucibles of innovation and creativity that promise to offer new solutions for resolving intransigent problems and challenges, solutions that probably would not emerge from within resource management bureaucracies simply because the bureaucratic organizational culture does not promote innovation and experimentation (Brick 1998).

Valuing community forestry initiatives as sites of innovation and ideas that might be institutionalized more generally is consistent with the view of community forestry that many practitioners themselves hold: that community forestry supplements, operates within, and interacts in a symbiotic manner with the existing legal, political, and institutional frameworks that govern public land management (Weber 2000). This in itself dispels some of the criticism that opponents level at community forestry. For example, sustainable community forestry programs and practices must meet the legal requirements concerning forest planning and management activities contained in all relevant legislation such as the National Forest Management Act and the Endangered Species Act. Actions that abrogate any of these laws are subject to legal challenge or agency refusal to participate in the proposed action. Community forestry activities must be at least as environmentally benign as any other allowable activity on public lands.[19] Similarly, community forestry practitioners do not seek to exclude the involvement of nonlocal communities of interest and national interest groups. Rather, they seek a better integration of local and nonlocal interests, one that empowers local voices that in the past have not been heard.

By developing supportive relationships with exemplary models of community forestry process and outcome, national environmental organizations could use this as a strategic opportunity to find ways to shed their elitist mantle and develop important alliances and coalitions with rural grassroots-based community organizations.[20] Indeed, some argue that the popularity of the "wise use" movement is a clarion call to the environmental movement to reach out to and create alliances with rural grassroots communities, people of color, and lower classes. A path of engagement, rather than opposition, would help to weaken the traditional antipathy toward some national environmental organizations in rural areas; equally or more importantly, it could create a potent melding of national and local forces based on the shared values of community, place, and sustainability (Hess 1996; Brick 1998; Rasker and Roush 1996).

The Forest Industry and Community Forestry

Although a full analysis of the forest industry as it relates to community forestry is beyond the scope of this project, a few key issues are highlighted here. Some of the important issues concern land tenure and ownership patterns, the scale of forest management operations, and issues of competitiveness and economics of scale. Not surprisingly, many of these issues are region specific. However, from a broad vantage point community forestry does not seem to register very prominently on the forest products industry radar screen. As one community forestry practitioner described when asked about the relationship between the forest industry and community forestry, "There is no relationship"; the industry considers community forestry "small potatoes." This person went on to suggest that the forest industry has not taken responsibility for its effects on land, communities, and the workforce and that although the industry has been involved in some small-scale community-based initiatives, it could and should play a larger role in terms of providing startup capital for community forestry organizations and in research and technology for community forestry.

Shifting focus from the national policy scale to a smaller scale at which the practice of community forestry comes into focus, it is apparent that corporate forestland ownerships raise diverse and challenging issues that vary dramatically from region to region. Some of the key issues concern access for customary activities such as subsistence hunting and fishing, access for subsistence and commercial nontimber forest products harvesting, and the working conditions of those who are directly employed by the landowner or whom contractors employ. Conner Bailey has noted that states such as Louisiana, Alabama, Arkansas, and Mississippi have become the woodbasket of the nation; 77 percent of the national pulpwood production now comes from the South. In recent years forest product corporations have been consolidating their forestland ownerships. For example, International Paper now owns 1 million of Alabama's 22 million acres of forestlands. Institutional investors such as John Hancock are also buying large forest tracts. These changes in forestland ownership negatively affect the potential for community forestry. Bailey points out that to attract forest industries, economically depressed rural counties offer substantial tax reductions. This reduces funding levels for local schools and other public services; in some counties in this region school budgets are one third what they would be if tax incentives had not been offered to corporate landowners. Enjoying a local tax burden as low as $1/acre/year, corporate landowners lease large tracts of land to hunting clubs for approximately $3/acre/year. As part of the lease agreement, hunting clubs also help maintain access roads, erect "no trespassing" signs, and generally secure the borders of the leased area to prevent poaching of "their" wildlife. These practices have effectively eliminated the rich array of customary subsistence hunting and fishing practices that local communities (African American, in most cases) have exercised on these lands for generations.[21] In some regions, the strength and importance of these customary claims were such that specific forest areas, such as fertile bottomlands, were managed to provide habitat for bear and deer that were hunted during annual clan and family-based gatherings.

The concentration of power through the consolidation of corporate forestland ownership can also have regressive impacts on forest workers. Much of the work on industrial ownerships is contracted out to large contractors who move crews of workers across large distances from job to job. In the South and much of the Pacific Northwest, many migrant workers are Hispanic. Questions have arisen regarding the working conditions for these crews, especially with regard to fair compensation and pay for overtime work. A class action lawsuit has recently been brought against Georgia-Pacific, International Paper, and Champion in Arkansas on behalf of this group of forest workers (Greenhouse 2001). The suit claims widespread violation of labor laws, including recordkeeping and wage and overtime laws. One of the linchpins of the suit is the argument that landowners are responsible for the working conditions of the contractors' work crews—that contractor employees, from the standpoint of labor laws and culpability, should be considered employees of the landowner.

A similar debate concerns the contractual arrangements under which brush harvesters gather wild floral greens on public and private lands in Washington State. A recent bulletin of the Jefferson Center for Education and Research (Umholtz and Brown 2002) focuses on this issue and provides the basis for the rest of this paragraph. The wild floral greens industry in Washington State generates hundreds of millions of dollars in revenue and involves thousands of brush harvesters. Although there are diverse contractual mechanisms for brush harvesting, an approach commonly used by larger brush companies is to purchase brush harvesting leases from timber companies or public forest agencies and then sell permits for harvest rights to a portion of the leased area to individual harvesters or groups of harvesters. The current controversy focuses on whether or not brush harvesters are employees of the brush companies who provide harvesters with subcontracts and permits for harvesting brush and who generally purchase from harvesters the brush they collect. At stake are questions concerning the applicability of Washington State labor laws, responsibility for providing harvesters workers' compensation insurance, and Social Security and other tax issues. The issue has assumed increasing importance because of rapid industry growth, concerns regarding the working conditions of large numbers of floral greens harvesters, and the possibility that low market prices due to an oversupply of floral greens on the market are fueling increased (and unpermitted) harvesting at ecologically unsustainable levels. During the last 2 years the Washington State Department of Labor and Industries and the U.S. Department of Labor have been holding a series of audits, meetings, and investigations to explore these issues. Meanwhile, a group of floral greens wholesale companies, who argue that harvesters are not employees, have asked the Washington State Superior Court for a summary judgment on the issue. The eventual resolution of the status of brush harvesters (whether they are employees or independent harvesters) will have important ramifications for other sectors of the contingent labor force, which is largely comprised of low-income and immigrant people, in the forestry as well as other sectors.

The issue of scale is also pertinent to this discussion of the intersection of community forestry and corporate forestry. Although small scale is not necessarily more equitable than large scale, it is definitely a more human scale, and typically, but not always, small contractors tend to be more a part of the community in which they live. Community forestry tends to be small scale, with small-scale contractors and crews, small scale in terms of forest acreage that can be quickly operated on (e.g., planted and thinned), and small scale in terms of the emphasis on business development using small-diameter timber from forest thinning and restoration. Small-scale community-based contractors and work crews have difficulty competing against large contractors whose economies of scale often allow them to underbid smaller contractors for work on both public and private forestlands. When Max Cordova, a community forestry leader in Las Truchas, New Mexico, asked the Forest Service contracting officer how he should modify his bids to successfully compete with nonlocal contractors for work on the Carson and Santa Fe National Forests, he was told he needed to cut his bid price in half to be competitive. Some also argue that the scale limitations of most community-based contract work are a serious disadvantage given the forest acreage in the western United States that needs forest restoration and fuels reduction work. The argument is that only an industrial-scale approach can mobilize the resources and people needed to address these large-scale forest management challenges. The notion of industrial-scale forest restoration work can raise red flags about the implications of this model for worker-related issues. Small-scale forestland ownerships, nonindustrial and industrial alike, are also more vulnerable to fluctuating timber and nontimber forest product prices. As free trade expands, forestland owners will become increasingly vulnerable to changes, especially decreases, in timber prices related to the increasing market availability of less expensive imported wood.[22] Larger-scale forest industry corporations often own forestlands in multiple countries; this buffers vulnerability to a depressed market in any one country (and in some cases enables them to take advantage of region-specific market depressions). Depressed timber prices, in part caused by the nonrenewal of the U.S.–Canadian softwood agreement and the subsequent increase in Canadian softwood imports, also threaten the economic viability of the rural businesses that are springing up around the sustainable use and processing of small-diameter wood.

In short, the relationship between the forest industry and community forestry is complex and multifaceted and varies from region to region. In some areas, it may be possible for community forestry to establish a "radical center" that will attract segments of both the environmental and forest industry communities. In some cases this is beginning to happen, but in others both environmental and industry groups still oppose community forestry as something that is too threatening to the status quo.

Conclusions

This chapter has provided a broad overview of the political, institutional, and government aspects of community forestry. The discussion of the implications of community forestry for the structure and organization of government parallels the discussion of civic republicanism in Chapter 6. The empowerment of communities and workers requires a new organizational model for natural resource management agencies and for organizing and prioritizing forestry extension activities. Drawing on both the practical experience of innovators and the observations of researchers, this chapter has sketched out some of the elements of that model. This chapter has also addressed the national-level political life of community forestry in terms of its own forays into interest group politics and its relations with the national environmental groups and the forest industry, the two primary interest groups with which community forestry intersects. Chapter 8 addresses the implications of community forestry and deep democracy for the practice of science and the production, ownership, and dissemination of knowledge.

CHAPTER 8

Toward a Civic Science for Community Forestry

(Jeffrey G. Borchers with Jonathan Kusel)

> The price of living in the world of the pragmatists and the skeptics is the need to acknowledge that our best-founded beliefs are still uncertain. Neither physics nor psychology can do what the rationalists hoped. Dreamers tempt us with their images, but only as poetry. When the dreams of theory no longer cloud our expectations, we are back in a world of practical hopes and fears.
>
> —*Toulmin (2001:204)*

Community forestry offers a powerful critique of the dominant model of knowledge creation and acquisition, which is rooted in the Progressive Era model of technocratic bureaucracies and scientific expertise. In contrast to this traditional model of science, community forestry espouses a participatory model of knowledge production, one that integrates monitoring and adaptive learning, incorporates local knowledge, and helps to empower groups. This chapter explores these themes by examining some of the underlying inadequacies in traditional scientific approaches to achieving the biophysical and social goals of community forestry. It describes a newly evolving, pluralistic model of science that provides a more flexible approach to integrating diverse epistemologies,[1] sources of knowledge, beliefs, and values.

Traditional forestry has evolved slowly in response to this and other radical social and technological changes that have overtaken the very science and society that once wholeheartedly informed and supported it. By clinging too long to the paradigms of centralized authority, mass production, efficiency, and the primacy of expert knowledge, Progressive Era forestry has, in effect, oversimplified its landscapes and its people. Following the lead of Gifford Pinchot and ignoring the voice of progressives such as Benton MacKaye (who advocated a broad and inclusive social agenda) traditional forestry saw a growing trend toward homogenization of viewpoints that ultimately created a profession poorly equipped to keep pace with the times, much less provide leadership to government and society.

Progressive and Modern: The Scientific Foundations of Traditional Forestry

What is the nature of traditional or Progressive Era science? This question has many answers and has long provided grist for the mills of philosophers, sociologists, and historians of science. Traditional science consists largely of a set of norms, or prescriptions, for objective learning, including paradigms (Kuhn 1962).[2] Although these standards for knowledge creation have evolved, their current form is inextricably linked to the science and mindset during the rise of technology and western industrialized nations in the nineteenth century, though with roots extending to seventeenth-century science. And although the Progressive Era may have given way to the modern and the postmodern, the traditional science of the late nineteenth and early twentieth centuries remains alive and well today.

Science historians and philosophers have long searched for the essential ingredients of science that distinguish it from superstition, indigenous knowledge, and pseudoscience. For example, in the eyes of many people, science remains the sole provider of objective standards by which social policy can be formulated. But to extend this essentialist argument further, if the ultimate source of science's power is intrinsic to itself, then its position of authority must reflect a superior way of knowing and believing. In such a philosophy, there is scant motivation for preserving or using other, "intrinsically weaker" knowledge systems. These essentialist arguments for the uniqueness of science have largely fallen into disfavor. This means simply that science retains its cognitive authority in society not so much by how it is practiced but how it is represented to society.

In recent decades, a critical reexamination of the fundamental tenets of science, more baldly called the "science wars" (Gould 2000), has revealed that traditional science's epistemology is largely illusory or unknowable. The result has been profound. Toulmin (2001:3) states, "The sudden loss of confidence in our traditional ideas about rationality in the last twenty or thirty years is marked enough, and widespread enough, to constitute . . . an episode, not just a collection of contemporary events: many writers today refer to it as the End of Modernity." The reexamination has led to a post-modern challenge to science's conferred authority and a questioning of its vaunted autonomy and objectivity and to questions about whether its search for truth has been co-opted by powerful interest groups. Its "pragmatic ends" are seen increasingly for their accompanying hidden costs in the form of undesirable social, economic, and ecological losses. Discussing decision making by the technocrats and the relationship of the exclusivity of their work to the decline of democracy, Fischer (2001:7) states, "Not only do experts lack answers to the complex technical questions that confront us, but expertise itself turns out not to be the neutral, objective phenomenon that it has purported to be." Struggling with local gridlock and buffeted by interest group "science," numerous community forestry practitioners now refer to scientists as an interest group.

Traditional Progressive Era forestry is a prime example of a profession's core principles caught in the crossfire of the science wars. As the true cost of nearly a century of traditional forestry has become apparent, its fundamental tenets—control, predictability, sustained yield, and maximization of production—have also been challenged. For the most part, this challenge to the scientific principles of forestry has been successful: The burden of proof has shifted from those who would question the science underlying Progressive Era forestry back to the profession itself. Rapidly changing social values, as exemplified by the rise of environmentalism and multiculturalism, expanded the entire frame of reference for a profession that, until the 1960s, bore a striking resemblance to agriculture. The rise of environmentalism and growth of ecology as a unifying science for natural resource management also cast a harsh light on the narrow goals, practices, and environmental consequences of industrial production forestry (McIntosh 1986). Although the results have been at times chaotic, the ensuing negotiations have created numerous opportunities to update the scientific foundations of forestry, incorporate a greater diversity of social values, and experiment with alternative ways of knowing and learning. Science is changing rapidly, and that change is forcing a fundamental reexamination of the legacies that traditional science has bequeathed to both communities and the land.

Restoring Communities and Forest Ecosystems: How Can Science Help?

As the twenty-first century opens, we face seemingly insoluble problems in society and the environment. In the not-so-distant past, the partnership between science and forestry promised rapid technological solutions to difficult but hardly insurmountable challenges. For instance, maximizing profits from forests managed for sustained yield of timber was a daunting task, particularly with erratic economic cycles. In such a restricted context, traditional science provided adequate guidance and continues to do so today. However, traditional science does not fare so well in more uncertain domains. If the sustained yield forest problem is placed in a more appropriate and larger context—the ecosystem encompassing diverse forms of natural and community capital—a number of high-stakes, high-risk social and environmental problems arise. These problems exhibit a high degree of system complexity and scientific uncertainty that render expert-driven solutions highly unlikely (Funtowicz and Ravetz 1993; Bradshaw and Borchers 2000). These are the "wicked" problems (Shindler and Cramer 1999) familiar to many: global warming, declining biodiversity, deforestation, pollution, and increasing joblessness and rural impoverishment, among others. In such ambiguous situations, traditional science hardly knows which questions to ask, much less how to proffer solutions.

From the perspective of forest-dependent communities and practitioners, wicked problems lose their abstract, global qualities and take on a harsher reality. The structure and composition of forest ecosystems—the natural capital—have direct and immediate consequences for the well-being of a dependent community. Conversely, the state of a community's physical, financial, human, cultural, and social capital can be invested in to restore or maintain natural capital. Restoring this relationship and maintaining "healthy flows" between community and natural capital is one of the principal goals of community forestry and restoration ecology (Higgs 1997) and may help define and achieve sustainability at local levels. However, it is an exceedingly complex and ill-defined goal that defies traditional approaches to management and problem solving. The challenge of creating a successful community forestry along these lines is, in short, a wicked problem.

Uncertainty and Complexity in Social and Ecological Systems

Is the goal of creating "healthy flows" between communities and the ecosystems on which they depend a realistic one? An answer to this question will be proposed later, but for now some of the intrinsic complexities of social and ecological systems that can impede progress toward this goal are discussed. The approach is motivated by the fact that the very same uncertainty and complexity that challenge traditional science and traditional forestry also represent obstacles to community forestry.

The task of restoring community and natural capital is a challenge involving not only the uncertainties that attend complex ecological problems but also the social complexities that underlie them. This is precisely the lesson that land managers are learning today, and it relates directly to the limited perspective of the science and forestry that imbue their thinking. For example, the Clinton administration's Northwest Forest Plan (USDA and USDI 1994), encompassing western Washington and Oregon and six northern California counties, represents a regional strategy for maintaining the viability of the northern spotted owl and other species within forest landscapes managed for a range of social and economic benefits. As a means for science-based policy formulation, the plan dealt reasonably well with the high levels of scientific uncertainty associated with future conditions of populations, landscapes, and economies. However, the Northwest Forest Plan and assessments like it are not a panacea for complex biophysical and social problems. Implicit in these efforts is the belief—indeed the assertion—both buttressed and relied on by political commitments, that the practice of traditional science will yield the understanding and answers necessary to provide timely solutions to difficult questions. Among scientists unmoored to the Progressive Era ideal of predictive control or enchanted by the "gospel of efficiency," there is a slowly growing recognition that science is incapable of providing the certainty desired by society and policy makers. Change is needed, and there is increased acceptance of this fact. In addition to the "failure" of scientists as autonomous experts, two primary developments drive this change in thinking. First, there are the recent developments in complex systems theory (Cowan et al. 1994), a multidisciplinary approach to describing the complex and unpredictable behavior of human and natural systems. Second is the rise of participatory research, a mode of inquiry that embraces public participation and deliberation as a part of research. The need for participatory modes of research is fueled by the complexity associated with problem solving, the need for new "contextually situated" information, and a growing recognition that the widening separation between the public and Progressive Era "experts" threatens democracy (Fischer 1990, 2001; Gaventa 1993).[3]

Complex systems theories have had a major impact on the ecological sciences, particularly ecosystem ecology. A complex system has a number of attributes (Cilliers 1998) and, as will be seen in the following description, may be relevant for almost any but the most simplified social or ecological system:

Many interacting elements: Forest ecosystems and local economies contain innumerable interacting parts. No model or plan can possibly account for them all.

Nonlinearity: Small causes may lead to large effects, or vice versa. For example, one road culvert plugged during postfire erosion may produce a major slope failure and sedimentation into a stream, thereby greatly reducing its value as spawning habitat for anadromous fish.

Feedback loops: Managers who previously installed road culverts observe that slope failure occurs when culvert diameters are too small. By installing larger culverts in burned-over areas, they nonlinearly decrease the rate of subsequent events.

Weak boundaries (hierarchy): Although fish may have distinct boundaries, streams do not. Their edges are diffuse, and they are nested within other large, weakly bounded ecosystems (e.g., watersheds, rivers, landscapes).

Far from equilibrium: A system at equilibrium is effectively an unwound clock. Living systems such as streams are like a clock with pendulum poised on the upswing. A stream ecosystem continually draws energy from external sources (e.g., water cycle, plants). This energy not only fuels the basic functions of a stream but also creates unique structures.

Emergent properties: Not every small culvert produces slope failure. The exact location of slope failures on a burned-over forest watershed cannot be predicted because of the large number of interacting parts acting nonlinearly through numerous unidentified feedback loops. The pattern of slope failures emerges as the system is observed.

Self-organizing: After rainfall, clogged culverts, slope failures, and stream sedimentation (a closer-to-equilibrium condition), the previous patterns and structures in soil and stream begin to reestablish as energy becomes available.

Unpredictability: Although it can never be proven, predictions about the future behavior of complex systems such as forest watersheds can never exceed a given level of accuracy or precision. For example, the pattern of reestablishing vegetation on a burned-over forest watershed can be only approximated at a large scale. For the forester who manages at small scales, predicting the emerging patterns may be a hopeless task.

In large part, the implication of complex systems theory for natural resource managers is embodied in the last point about unpredictability, particularly when stakes are high. The inherent unpredictability of complex systems—such as forest ecosystems and community economies—translates to a failure of traditional styles of management, decision making, and policy formulation (Gunderson et al. 1995; Gunderson and Holling 2001). To a large extent, this is a failure of science and its underlying epistemology that inform those activities. Yet science, and scientists, continues to evolve, driven by a growing backlog of urgent social and environmental problems and by criticisms of its culture and epistemology (Nader 1996; Ahl and Allen 1996; Berkes 1999). Although complex systems theory may undermine some of the comfortable paradigms of traditional science, it is important to point out that it was traditional science itself that led to the new theory. Science, as part of the complex social system, is being transformed along with society and the environment, sometimes in unpredictable ways. Because many citizens are less concerned with the sanctity of science's knowledge base and more with taking the precautions necessary for avoiding undesirable risks (Francis and Shotton 1997), a rejection of traditional science does not represent a rejection of traditional science outright but reserves it as a tool to use in more circumscribed contexts.

Beyond Traditional Science, Inclusively

As we discussed previously, the wicked social and ecological problems are characterized by uncertainty and high stakes. These complex systems—which include a diverse array of knowledge types and social values—can be confronted only as science begins to evolve and accommodate the new challenges. What ultimately will emerge in this process reflects a desire to rethink some critical, unquestioned assumptions of traditional science. The critical distinction of a posttraditional science is that it rejects the illusion of predictability and control that characterizes traditional science and rejects also exclusive reliance on the expert. It substitutes in its place a concrete agenda for accommodating the "radical uncertainty and plurality of legitimate perspectives" (Funtowicz et al. 1999:7) concerning complex social and ecological systems.

Participatory and collaborative research strategies are both a response to the failure of traditional science and part of a posttraditional science involving a plurality of legitimate perspectives. These strategies have come into focus because of their success outside the United States, particularly in developing countries and involving the poor (Chambers 1997; Tandon 1998), and because of the promise of public involvement offered through various state and federal planning requirements and law (such as the National Environmental Policy Act and the National Forest Management Act, to mention just a couple). Community forestry in the western United States has been jump-started as a result of public participation opportunities associated with federal land management, even though participation has been highly rigid and constrained. The rise of participatory and collaborative research in natural resources stems also from an increasing activist stance on the part of people affected (and frustrated) by resource management who want more of a say in decisions that affect them and who want to take more responsibility for strengthening civil society, improving their own understanding and condition, and redressing imbalances of power.

Public involvement in the practice of science through participatory research raises the inevitable questions: What constitutes knowledge? How is the knowledge of the public to be integrated with that of the scientist? We respond to the second question later in a discussion of adaptive management. In response to the first question, the movement beyond a traditional science based on knowledge that is abstract, quantitative, statistically valid, and generalizable creates space for knowledge based on *practice,* about which community forestry practitioners have much to say. Knowledge based on practice is located in place and set in the context of time and historical circumstances. This shift means that no longer is practice subservient to theory, yet neither is it superior. Toulmin (2001:171–172) calls this restoration of balance between the two a "return to reason." The challenge remains, however, in reconciling the two through the practice of deliberation. Deliberation involves a process of civic discovery and mutual learning that, according to Robert Reich (cited in Fischer 2001:226), focuses on "how problems are defined and understood, what the range of possible solutions might be, and who should have the responsibility for solving them." Describing participatory research in the context of community inquiry and local knowledge, Fischer (2001:191) states,

> Politically, participatory research's dedication to democratic practices provides a dramatic departure from the mainstream commitment to the corporate-bureaucratic state. On the epistemological level, its emphasis on collaborative research and the methodologies of problem posing, discourse, and social learning confront the most pressing and sophisticated epistemological issues facing the social sciences.

To return to the issue raised in our previous discussion, is the goal of creating "healthy flows" between communities and the ecosystems on which they depend a realistic one? Not too many years ago, the answer from traditional science would have been an emphatic "yes." However, posttraditional science, with its greater appreciation for complexity and uncertainty, answers "maybe" and sets out some conditions for improving the likelihood of success of the project. In the next section, deliberative and integrative processes for engaging science and scientists are described. These approaches do not guarantee success in community forestry; however, they do favor the emergence of a sustainable relationship between communities and ecosystems in that they accommodate diverse social values and sources of knowledge while maintaining a learning environment.

Civic Science Partnerships and Adaptive Management

One of the better-known models for pursuing the goals of community forestry is adaptive management (Lee 1993; Walters 1986; Holling 1978). Adaptive management is deceptively simple in its recommendations to assess, design, implement, monitor, evaluate, and adjust (B.C. Forest Service 1999). Monitoring the social and ecological outcomes of management actions (or inactions) is a crucial feature of adaptive management, for it allows one to determine whether management goals have been achieved. Effective monitoring also means that the inevitable surprises associated with managing complex systems will be occasions for learning rather than lamentations over failure. More active forms of adaptive management increase the rate of learning by designing and implementing landscape-scale experiments that explicitly test management hypotheses (Walters and Holling 1990).

Despite the many desirable features of adaptive management, there are few examples of its successful implementation. This is perhaps because its principles are poorly understood and difficult to apply in the face of numerous social, scientific, and institutional barriers (Kusel et al. 1996). For example, science generally provides land managers with only indirect guidance in the form of technology transfer, applied research, and consultation. Generally there are few, if any, opportunities for the two-way exchange and deliberations necessary for creating informed partnerships. This traditional pattern still holds, even though adaptive management considers direct collaboration between managers and scientists a prerequisite for success (Walters and Holling 1990). Overall, science remains largely detached from efforts to build collaborations between the public and land managers. These barriers compel public agencies such as the Forest Service to play the difficult role of intermediary between science and the public. A legacy of adversarial relationships often plagues efforts to bridge communication gaps between science, the public, and management. When environmental issues are complex and ambiguous, the public, science, and agencies compete legally and politically to influence policy formulation and decision making. Although this competitive process may contribute to a dialogue, more often than not it also means that many pressing social and environmental issues may not be addressed in a timely fashion.

In light of these obstacles, as well as the inherent complexity of managed ecosystems, community forestry needs a well-designed process of adaptive management to realize its mission of restoring natural and community capital. However, implementing adaptive management is a complex undertaking, one that takes substantial work at the boundaries of science and land management agencies. Lee (1993:161) endorses adaptive management for such tasks but goes on to say that "managing large ecosystems should rely not merely on science, but on *civic* science; it should be irreducibly public in the way responsibilities are exercised, intrinsically technical, and open to learning from errors and profiting from successes."

Civic science involves a blending of science and politics, using a full spectrum of players concerned to varying extents with "truth" and "power." In a civic science partnership (Figure 8.1), scientists, politicians, administrators, and public have different allegiances to these conflicting goals. For example, whereas scientists are concerned mainly with contributing to a

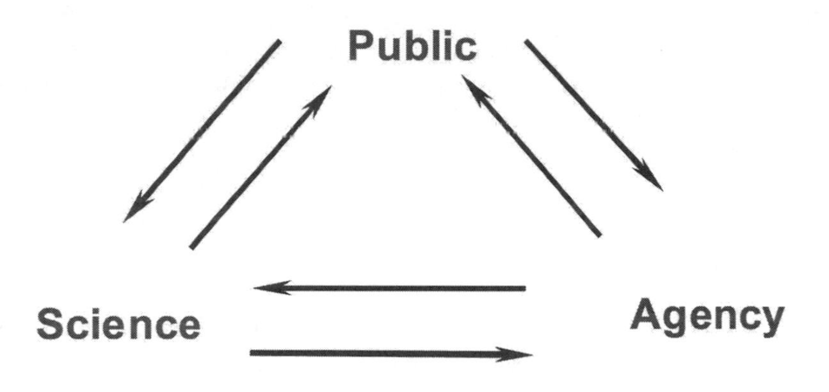

FIGURE 8.1 The Public–Science–Agency Partnership Model

consensus view of truth, land management agencies, in their exercise of power, are accountable to voters via elected officials. Where are the communities and workers in the civic science scheme, and what can they contribute? Both residents and workers can engage with scientists, wielding local knowledge and a diverse set of values. Local residents and workers who are well informed about the perspectives of scientists on issues they are concerned about can more effectively advance their own knowledge and values. Conversely, scientists who are well informed about community knowledge and values are better positioned to facilitate adaptive management experiments and research that have greater meaning and are more responsive to the community and society at large.

Although collaborations between the public, science, and agencies are essential to the success of adaptive management, there is more. Collaboration is a necessary precondition for creating an environment conducive to learning (Parson and Clark 1995). In adaptive management, learning is envisioned as a viable alternative to crisis management. Yet given the high levels of complexity, uncertainty, and conflict surrounding most environmental issues today, achieving the high standards for learning under adaptive management entails the judicious design and application of methods and processes that will meet the learning expectations of all parties. Rooted in an epistemology of posttraditional science, these methods embody two important goals for a civic science partnership: Ask the right questions (Marcot 1998), and answer the questions right.

Ask the Right Questions

The first goal has important implications for the science–public relationship because the lessons learned by scientists and resource managers often are biased by the questions they fail to ask. The struggle to formulate appropriate research questions and derive well-structured hypotheses is the foundation of a good learning strategy in adaptive management. Yet the privilege and power of formulating strategic questions is traditionally vested with professionals—scientists, managers, and policy makers—whose values, knowledge, and biases ultimately shape the learning process.

For environmental issues such as sustainability, the process of formulating strategic questions must include more diverse and representative perspectives that can ensure that the right questions are asked. Sustainability is a prime example of knowledge problems that remain ambiguous or equivocal (Zack 1999). From the standpoint of adaptive management, the principal challenge of sustainability is not so much complexity or uncertainty but articulating or selecting appropriate questions for research and learning. Fischer (2001:250) states, "The goal of discourse is not just 'to improve understanding' but rather to create it through exploring the social meaning of the research and its findings." In this realm, the methods of posttraditional science are more useful as an adjunct to a democratic process that encourages a diversity of viewpoints.

In a civic science partnership, this democratic process starts locally. The inclusion of communities as participants in formulating strategic questions produces a scope of inquiry relevant for local people and workers. A community's need for knowledge often reflects an interest in local sustainability and in increasing its capacity to anticipate and adapt to the forces of social, ecological, and economic change. Ultimately, communities and their surrounding landscapes make up the fine-scale building blocks of sustainability. Their relationships, values, and knowledge of place provide an essential foundation for building social, ecological, and economic sustainability at bioregional and larger scales (Borchers 1996).

Answer the Questions Right

If learning begins by asking the right questions, then how best can we address those questions? The second goal of a civic science partnership is directed toward finding methods that can enable learning in the process of addressing strategic questions. The learning methods in science are particularly powerful as they attempt to ensure strong inferences about observations and experiments. Strong inference in adaptive management means simply that the often costly process of monitoring provides an adequate test of the hypotheses concerning the outcomes of management actions (Lee 1999). This strength of inference depends mainly on wise planning: the quality of hypotheses, experimental designs, implementations, and analyses. If these qualities are present, then to the greatest extent possible learning is enabled, and areas of conflict and disagreement are more likely to be addressed effectively. This is critical because many of the controversial questions relating to sustainability can be articulated and addressed only in experimental settings where conflict can be "bounded" (Lee 1993, 1999). Strong inference therefore ensures that adaptive management experiments will provide the critically needed lessons that civic science partners define.

One way to strengthen inference in adaptive management is by the process of inclusive monitoring. By *inclusive monitoring* we mean an all-party monitoring process in which all stakeholders in the outcome of a proposed management activity participate. In addition to providing important feedback about land management practice, monitoring, coupled with an effective reporting system, is essential for establishing a system of checks and balances that protects interests and advances deliberative processes. In community forestry projects, all-party processes build trust and, when successful, provide quality assurance and accountability for a wide array of stakeholder groups (Gray et al. 2001; Bliss et al. 2001). Although inclusive monitoring stresses broad democratic participation, it does not preclude the use of well-articulated, testable hypotheses, statistically robust experimental designs, and rigorous implementations and analyses to strengthen inference. These are the well-tested core competencies of traditional science: various methods, tools, and technologies that, under certain conditions, may serve as a benchmark for diverse stakeholder groups wanting to test different landscape hypotheses within a posttraditional, adaptive management framework. In instances of overwhelming system complexity, and particularly one with shifting values, traditional science may not deliver the level of certainty and clarity desired by stakeholders engaged in monitoring. In such cases traditional science may be relegated to a less dominant supportive role, where its powerful toolbox serves not as a path to objective truth but as a way to enhance the quality of learning in inclusive monitoring.

A Civic Science Strategy for Community Forestry

Designing a civic science approach to adaptive management and community forestry is a complex undertaking. Any strategy toward that end should give collaborative learning a high priority, particularly the critically needed lessons for restoring community and natural capital in forest ecosystems. This collaborative process involving communities, scientists, and managers should reflect a consensus about the roles that partners should play to better direct and facilitate learning.

Although collaboration and learning represent the ultimate goals of a civic science strategy, the means by which these can be achieved are numerous and various. For example, to build and maintain trust, a civic science partnership must first maintain transparency with regard to the rationales, motivation, assumptions, and agendas of its daily operations. With the greater trust afforded by such transparency, a supportive learning environment can lead to the type of exchanges between partners that are crucial to learning in adaptive management.

The democratic and epistemological roots of a civic science strategy suggest several broadly defined roles and functions for a civic science partnership and its members (Box 8.1). Although detailed aspects of these roles should be defined as part of a democratic process, there are several broad goals that civic science partners should consider.

Box 8.1 Example of Exchanges Between Civic Science Partners in Adaptive Management

- *Science to Public:* Provide decision support, education, relative negotiability of ecological constraints; interpret ambiguity, equivocality, complexity, and uncertainty.
- *Public to Science:* Provide research and monitoring objectives, hypotheses linked to community values and objectives, and local knowledge; convey learning expectations.
- *Agency to Public:* Define decision context, describe regulatory environment, enable monitoring, convey management agendas.
- *Science to Agency:* Provide scientific checks and balances and credibility for planning and decision making.

Communities as Knowledge Integrators

To become effective partners with scientists, communities—including residents and workers—should acquire a sense of ownership of science. As users of science, an appropriate role for communities is to provide strategic guidance (i.e., ask the right questions) in applying the methods, tools, and knowledge that science embodies. Thus, communities assume the role of knowledge integrators (Zack 1999) at a local level, a position that entails partaking not only of what science has to teach but also of its powers of investigation and decision making. By consulting with scientists and agencies in this capacity, communities can more effectively render their own knowledge and values into adaptive management hypotheses and experiments. As partners seeking to define the right questions, communities, whether they are a community-based watershed or forestry group in the West or a cooperative of small forestland owners in the Midwest, can also provide leadership, particularly in ambiguous areas such as restoring forest ecosystem health.

Scientists as Learning Facilitators

Beyond its traditional educational role, what can science do for a civic science partnership? In her call for a new social contract for science, Lubchenco (1998) asks researchers to "construct more effective bridges between policy, management, and science, as well as between the public and private sectors." For scientists in a civic science partnership, this means that scientific paradigms and models must accommodate community knowledge and belief systems. Scientists should provide community users with an understanding of the limits of scientific knowledge and the effects of ignorance and uncertainty on its predictions (Morgan and Henrion 1990). They can also portray honestly to communities the values that underlie scientific understanding, particularly in the type of questions it does and does not ask. Finally, science can ensure that strong inference and other rigorous learning strategies are safeguarded in adaptive management experiments.

Agencies as Learning Guardians

Managers in agencies who have decision authority over public or private lands often attempt—with limited success—to represent the competing values and knowledge systems of science and the public. In a civic science partnership, however, agency professionals assume a more appropriate role by creating and maintaining a viable context for learning. This may include acting as a clearinghouse for data, information, and knowledge acquisition; integrating and implementing adaptive management experiments into larger-scale plans and projects at a watershed or larger ecosystem level; and identifying management approaches and monitoring plans that are consistent with environmental regulations. For example, with sufficient financial, logistical, technical, and moral support from resource management agencies, the 10 adaptive management areas established by the Northwest Forest Plan can provide an ideal context for designing, creating, and testing a civic science strategy. As guardians of the infrastructure of adaptive management, land management agencies are in the best position to nurture a potentially fragile exercise in democracy and to ensure a fair and adequate evaluation of its goals and achievements.

The Civic Science Partnership as Strategy Maker

Although a civic science partnership is likely to have many unexpected, emergent properties, it should provide that which no single partner can reliably provide: learning strategies for sustainability. For example, by applying and testing various criteria and indicators for ecological, social, and economic sustainability, much of the ambiguity and disagreement about human interactions with natural systems can be resolved. This requires a strategy of "embracing uncertainty" (Anderson 1998), not only as a way to achieve management objectives for the landscape but also to identify and fill critical knowledge gaps. In adaptive management experiments, there is pervasive uncertainty about the tradeoff between landscape management objectives and knowledge that must be acquired. Paradoxically, learning the critically needed lessons of forest restoration may entail bolder, perhaps controversial experimental approaches (Carpenter et al. 1999). If it chooses, a civic science partnership is well suited to engage in such learning strategies, particularly as it represents a broad and diverse social commitment to use adaptive management as a means for resolving areas of ambiguity and controversy.

Obstacles to Civic Science and Adaptive Management

A civic science approach to adaptive management is difficult to put into practice. Translating adaptive management into on-the-ground practices has proven difficult for scientists and managers alike. The obstacles derive from many sources, but in U.S. land management agencies, the most noticeable are a lack of opportunities for meaningful public participation and a lack of decision-making transparency. A recent report by the General Accounting Office (GAO 1997:7) reviewing the U.S. Forest Service's decision-making capabilities states,

> Long-standing deficiencies within the agency's decision-making process, which have driven up costs and time and/or driven down the ability to achieve planned objectives, have not been corrected. These deficiencies include (1) not adequately monitoring the effects of past management decisions to more accurately estimate the effects of similar future decisions and to modify decisions when new information is uncovered or when preexisting monitoring thresholds are crossed, (2) not maintaining comparable environmental and socioeconomic data that are useful and easily accessible to forest managers, and (3) not adequately involving the public at the beginning of the decision

making process when problems are identified, data are gathered, and relationships are established and maintaining their involvement throughout the process.

According to a recent policy analysis, this legacy has hampered efforts to implement adaptive management: "The innovative and aggressive programs required to deliver on the promise of adaptive management might be seen as not worth the investment or risk, especially when public support and political will are problematic" (Pipkin 1998:70). As a result, the flexibility and adaptability of adaptive management as envisioned by the Northwest Forest Plan (USDA and USDI 1994) have failed to materialize, and "advocates of both environmental enhancements and commodity outputs see their role as protecting their interests from further erosion" (Pipkin 1998:70).

This suggests that even intuitively appealing methods such as adaptive management cannot be easily imposed from the top down. Environmental conflict coupled with complexity, uncertainty, and risk implies that truly adaptive management responses must be the byproduct of a civic science process for making high-quality, defensible decisions. Adaptive management provides some strategic guidance in this task, but its philosophy remains insufficiently articulated to provide planners and resource specialists with specific strategies and tactics.

Monitoring in the context of adaptive management also remains an unresolved conundrum, in part because it must fulfill so many diverse expectations. For the public, monitoring serves as a watchdog on ecosystem health, management, and policy. For the scientific community, monitoring provides critical scientific insights that ultimately inform the public, policy makers, and decision makers. In addition to scientific recommendations, policy makers also derive indirect benefits from the assurances that monitoring provides to their constituency. Resource specialists and managers benefit from all these elements. However, there are few occasions when resources or time for monitoring suffice to fulfill the panoply of expectations.

From the standpoint of decision making within adaptive management, monitoring programs must determine what, how, and when to monitor communities and landscapes with the accuracy and precision required by science, management, law, and policy. Such issues are rarely straightforward. For example, the scales governing ecosystem functions typically are much larger than the short-term cycles of political tenure and funding. The resulting spatial or temporal undersampling means that much of the data collected in monitoring programs cannot withstand tests of statistical significance (Bradshaw 1998). Moreover, these problems of quality control are compounded by problems of quantity: Limited fiscal resources for monitoring constrain the rate of critical knowledge acquisition. For these reasons, many monitoring programs may yield little short-term insight in the long-term processes of social learning and adaptation that accompany landscape experiments (Walters and Holling 1990).

Conclusions

As a top-down approach, designing and creating a civic science strategy for adaptive management and community forestry are highly impractical. The lessons of complexity theory and the increased demand for participatory research suggest that many of the desirable features of civic science will have to emerge spontaneously out of hundreds of grassroots experiments currently taking place. This argues for a more systematic approach for catalyzing these processes and for retaining and replicating the many lessons learned. Science, particularly the posttraditional version, can provide a much-needed rigor to emergent civic science partnerships. The boundaries of science are becoming increasingly porous with respect to community forestry, and scientists are increasingly willing to experiment outside the narrow confines of disciplines and research institutions (Bradshaw 2001). There are many pragmatic reasons for scientists to engage in civic science partnerships, and in a democracy, scientists have little to fear from working with people who bear the costs and benefits of science. Finally, there are many reasons for land management agencies to pursue civic science partnerships. Although the loss of power and control of this approach may at first appear threatening, civic science can be empowering for agencies whose credibility has been diminished over the last several decades.

CHAPTER 9

Community-Scale Investment, Equity, and Social Justice

(Leah Wills with Mark Baker)

> Americans [have] not yet learned the difference between yield and loot.
>
> —*Sauer (1938/1963:154)*

Community forestry arose, in part, as an attempt to stem the flow of value from ecosystems and the communities whose well-being is tied to them by integrating investments for forest ecological restoration with opportunities for community revitalization. Part of the effort to build the basis for greater investment levels concerns identification of the full spectrum of benefits that forests provide. The role of forests in moderating global climate fluctuation, regulating atmospheric storage and release of carbon and greenhouse gases, and providing biological reserves for many of the nation's most threatened terrestrial and aquatic species during parts of their life cycles can be of immense benefit to community forestry. Air, water, climate, and biodiversity issues affect relationships between forest-dependent communities and the forests on which they depend. These larger issue policy effects can be positive or negative, sweeping or minor, depending on how community engagement and community–forest exchange issues are designed. Policies that recognize that forests provide essential public goods can strengthen community forestry if they constitute the basis for increased community-scale forest investment. Community forestry engages these larger issues as it simultaneously focuses on policies that more directly address the relationships between forests and local communities such as contracting and workforce policies, tax policies, and timber harvest levels.

With the help of the capitals framework, the assets maintenance approach, and insights from the concept of ecological poverty, this chapter analyzes forest investment from the perspective of ecosystem health and community well-being. We briefly review some of the reasons why more value flows out of ecosystems and communities than back in. This leads to a discussion of investment strategies that have been developed to promote reinvestment in forest ecosystems. Although many of these strategies have begun to be implemented and many of the institutional changes necessary to actualize them are already occurring, the links between investment, equity, and community well-being often are not well developed. Therefore, it becomes difficult to achieve the interdependent community forestry objectives of reinvesting in forests and enhancing community well-being. Investment strategies that explicitly integrate environmental and community investments are needed. Although not always, such strategies often are linked to issues of social justice and equity. A case study that illustrates the potential of such integrated investments is presented. With a focus on both rural and urban contexts, it illustrates the potential linkages between community-scale reinvestment in forest ecosystem services and improvements in community well-being. The chapter then addresses more general issues concerning the challenges and opportunities associated with promoting community-scale forest ecosystem investment.

Disinvestment Drivers

Multiple factors—institutional, economic, political, and others—account for the gradual depletion of both natural and community capitals. Some of these factors are related directly to disinvestment. For example, the larger financial systems within which forest ecosystems are embedded tend to pull economic value out of the ecosystem. This is partly because rates of forest regeneration rarely match the rates of financial return that corporate shareholders want. Therefore, from a short-term profit-maximizing perspective, it makes sense to liquidate natural capital, convert it to financial capital, and invest it in non-forestry enterprises with higher rates of return. This leads to forest conversion or to forest depletion and maintenance in a simplified and fragmented condition with reduced capacity to provide the full suite of ecosystem products and services associated with forests (Best 2000:4). This tendency is exacerbated when high-interest junk bonds are used to raise the necessary short-term capital to purchase forestlands and associated operations; forest liquidation is the all-too-common method for paying off high-interest debt.

There are also more passive features of the institutional landscape that reinforce unidirectional flows of value. For example, market values exist for only a small proportion of the wide array of goods and services that forests provide. To date there are only limited mechanisms for compensating landowners for providing nonmarket forest ecosystem goods and services. The result is that forest management regimes focus on the production of forest products for which market values exist, thus underemphasizing the provision of the broader suite of benefits forests can provide (Luzadis et al. 2001; Best 2000). Economic systems also tend to pull value out of forest (and other) ecosystems because the values assigned to ecosystem products rarely reflect their full cost (social and environmental). Competition on global markets reinforces below-cost pricing of forest products. It can also pull products away from localities, thus diminishing opportunities for value-added processing and building financial capital.

From a conceptual standpoint, economists often do not distinguish between long-term natural capital maintenance and annual financial revenues and costs. This leads directly to underinvestment in capital maintenance activities. These are distinctly different kinds of analysis, although they are often conflated. Confusing asset maintenance with the balancing of current benefits and costs often leads to an overallocation of natural resources and an overconcentration of financial capital. One result is depleted natural capital.

Although there are few quantitative studies of disinvestment as it relates to the depletion of natural capital, an economic assessment was included with the biological and social parts of the Sierra Nevada Ecosystem Project. The Sierra Nevada Ecosystem Project's investment analysis can be used to illustrate the general magnitude of the disinvestment challenge. The Sierra Nevada Ecosystem produces direct resource values of $2.2 billion annually in commodities and services.[1] Only about 2 percent of these resource values are captured and reinvested in ecosystems or communities through taxation or revenue-sharing arrangements; this level of reinvestment is inadequate to ensure sustainable utilization of the ecosystem (Stewart 1996:974). Stewart proposes four reasons for the lack of investment. These include the valuation of ecosystem attributes in a manner that discourages investment, exchange restrictions that prevent the formation of value for ecosystem attributes that generate economic benefit, institutional barriers between agencies and government that prevent the capture of economic value where these are known, and a lack of capacity by localities to capture and reinvest the economic surpluses they generate for ecosystem health and community well-being.

Current and Proposed Investment Strategies

The patterns of forest disinvestment just described suggest that without some prodding, the market alone does not generate adequate levels of investment for natural capital asset maintenance. Although the market may allocate commodity resources efficiently, the resulting pattern of resource allocation may not be equitable or sustainable, especially when the broader range of forest attributes is considered. Recognizing these limitations of the market, economists such as Sen (1999) admonish their colleagues to avoid the all-too-common uncritical reliance on the market to determine the allocation of scarce resources. Sen also questions the widespread faith in Adam Smith's invisible hand, especially the assumption that "the ubiquitous selfishness" of individuals will automatically advance the interest of society (1999:118). Like other economists, Sen emphasizes the importance to capitalism of public values and goods. And he underscores the crucial importance of the value systems, mutual trust, and norms that underlie functioning capitalist markets. He argues that the two key challenges that capitalism must face are how to effectively address persistent inequality in a context of unparalleled economic growth and how to ensure environmental sustainability and the continued provision of public goods (1999:267).

The issue of how to maintain and sustain public goods has long occupied the attention of economists. Ecological economists have proposed that green taxes, depletion taxes, or other means of full social and environmental cost accounting and accountability be implemented to rebuild public goods. These market mechanisms support investment in and maintenance of natural capital. Other strategies such as public investments in new resource-efficient technologies are options that could add significantly to local well-being. The focus of ecological economists has been on increasing societal benefits and lowering social and environmental costs associated with forest commodity extraction in domestic and global markets. Identifying and harnessing opportunities for developing markets and market-like mechanisms that recognize a broad suite of forest values and reinvesting a portion of that value back into the forest ecosystem are a key concern. Such efforts focus on identifying and creating markets for the full spectrum of goods and services that biologically diverse and ecologically functional forests provide to create the means whereby forest landowners may be compensated for conserving and restoring their forest assets.

A variety of strategies are being developed (and some have been implemented) to curb disinvestment in natural and community capital and instead promote reinvestment. These strategies integrate emerging financial tools and organizational arrangements with the broader attributes of particular forest ecosystems. They attempt to provide financial incentives to forest landowners to encourage actions that conserve and restore forests and increase natural capital values. These emerging strategies have begun to attract significant levels of investment while promoting environmental and community values.

Following recent research on ecosystem reinvestment (particularly the work of Luzadis et al. 2001), the diverse array of forest investment strategies can be grouped into three broad categories.[2] The first category concerns the wide variety of private investment opportunities. This includes third-party forest certification, carbon sequestration packages, resource banks, partnership financing, and technological developments. Forest certification is a well-established market-based system that assures the consumer that the production and processing of forest products bearing the logo of the certifying organization are consistent with principles of responsible forestry as defined by the certifying body. Although the benefits of certification are diverse, one primary anticipated benefit is that consumers will be willing to pay more for certified wood, and forest owners and managers will be compensated for investing in responsible forestry practices through a higher return on their products.[3] Carbon sequestration packages are attempts to compensate forestland owners for the ecosystem services trees provide in terms of the atmospheric carbon they sequester. Resource or forest banks are institutions that purchase title to a forest while the landowner retains title to the land. The landowner receives a principal deposit in the bank and earns interest on the principal, while the holder of the forest title manages the forest according to certification or other region-specific criteria.[4] Partnership financing involves financing investments in equipment and machinery associated with low-impact certified forestry (especially at the community scale of nonindustrial private forestland ownerships) by banks and other financial institutions. A key element of partnership financing is the need to expand the science of low-impact harvest systems and ecologically grounded silviculture such as forest thinning and fuels reduction. Examples of relevant technological developments include those that enable the harvesting and milling of small-diameter logs to maximize the value-added attributes of ecosystem management byproducts. Expanding research and development of forest products (especially value-added products), quantifying and packaging the ecosystem services forests provide, and developing new and better markets for forest ecosystem goods and services are core elements of these strategies.

The idea of resource banks is related to the broader set of strategies such as the familiar conservation easements, in which a landowner sells some portion of his or her property rights (often the right to develop the land) to an organization that is bound to hold that right in trust, in exchange for financial incentives such as tax relief and sometimes direct payments. "Social easements" could be linked to conservation easements. They would secure the provision of various socially valued public goods, such as continued public access to footpaths and hiking trails or access to customary gathering areas for indigenous groups. Many of these private and philanthropic investment innovations are based on the assumption that property rights consist of a separable bundle of rights that can be packaged and exchanged in different combinations, providing that the mechanisms of exchange are developed and secure. Other investment innovations, such as partnership financing and technological advances, seek ways to add value to the outputs of sustainable forest management.

A second set of strategies Luzadis et al. review are those concerning efforts to realign political and social boundaries so as to create opportunities for exchange between those who value ecosystem management and those who manage, own, or work in ecosystems. Examples of some of these efforts include watershed organizations that link upstream land managers with downstream water users, zoning and tax programs that compensate landowners for maintaining the integrity of forest landscapes while benefiting from the revenue generated by higher tax rates outside forest zones, and coordination between adjacent counties (and state and federal government) to guide development pressure and create exchange mechanisms for compensating owners of areas for which nondevelopment land uses are considered a high priority.

The third strategy set concerns innovative forms of government financing that promote investment in ecosystem services. This set includes strategies such as making federal budgetary allocations for forest restoration work consistent and mandatory rather than inconsistent and discretionary, stewardship contracting on public lands, and international trade reform that protects producers who use sustainable but costly management practices from being undersold by other producers who produce forest products less expensively and less sustainably.

A common theme throughout most of these strategies is that restoring degraded and fragmented forest ecosystems and revitalizing forest-dependent communities take a comprehensive capital formation portfolio that incorporates a range of investments. A capital investment portfolio should reflect the specific needs of a particular forest region's ecological, social, and administrative character. Private, nonindustrial hardwood forests and their communities, urban forests and their communities, key public watersheds or recreational gateway forests and their communities, or special forest plant harvesting areas and their communities would need to develop different kinds of capital investment portfolios. A community and environment investment portfolio might include a mix of the following strategies, depending on the particular characteristics of the affected forests and the communities:

Investments for adding value to restoration forestry byproducts. Sustainable commercial wood production produces undervalued byproducts such as biomass and milling waste. Making commercial use of the small-diameter trees and nontimber vegetation that are removed for forest restoration purposes helps to underwrite the costs of forest restoration or broadens the economic benefits of forest restoration. There are many examples of this kind of investment, such as the Navajo Indigenous Enterprises housing project; ginseng management in the Appalachians; the New Mexico–based Forest Trust's Wood Products Brokerage, which serves family-owned logging and manufacturing businesses that produce *vigas* (ceiling beams) and *latillas* (sticks for ceiling material) for use in adobe construction in the Southwest; and the proliferation of family-owned and small-scale hardwood veneer operations in the Northeast.

Fair share maintenance. Distant beneficiaries would sustain the ecological function, structure, and species diversity for key watersheds and recreation forests through user fees or operation and maintenance fees. Watershed investment partnership programs are being piloted by the Los Angeles Department of Water and Power in the distant Mono Lake basin and by Seattle Power and Light, Montana Pacific Power and Light, and the Denver Water Board in their local watersheds. The Baker City (Oregon) Watershed National Stewardship Pilot Project is an example of an investment portfolio developed for federal forests that are also municipal watersheds.

Paying sooner rather than later. Reducing environmental impacts or enhancing the effectiveness of environmental mitigations could include investing in the ongoing development and testing of environmentally friendly technologies or environmentally friendly management practices. Fuels reduction work done with lower soil compaction harvesting equipment (even horse logging or hauling logs on frozen ground) and wood product development done with less wasteful processing equipment such as solar kilns are examples of these kinds of investments. Investments to avoid higher production and environmental costs in the future can be structured to also produce community benefits or at least to avoid disproportionate harm to communities.

Investments in scaling up. Securing market share may take increased worker, producer, and landowner cooperation for more efficient production or marketing. Fragmented forest ownerships and overconcentrated or widely dispersed markets often preclude effective participation by community-scale enterprises and nonindustrial forest owners. The Healthy Forests–Healthy Communities marketing network in the Pacific Northwest and Interior West, Michigan's Western Upper Peninsula Forest Improvement District, and nonindustrial forestland owner networks in the Northeast are examples of scaling up investment strategies.

Avoided cost investments. Investments in infrastructure alternatives that avoid long-term social and ecological costs are in the public interest, even if they cost more in the short term. TreePeople's Second Nature and Cool Schools projects in Los Angeles and the New York City watershed partnership are examples in which investments in community-scale green infrastructure are almost cost-competitive with investments in centralized "gray infrastructure" projects such as flood control works and water treatment plants but provide broader social and environmental benefits.

Incremental opportunity cost investments. These are a bundle of strategies aimed at investing in local economic diversity in ways that sustain a more diverse array of forest attributes and ecological services. Often, the fragmentation and conversion of forest environments have not occurred to a critical (irreversible) extent precisely because there has been some incremental level of economic return for conserving ecological integrity. Absent these incremental investments in ecological and economic diversity, the cost of saving the last few salmon, goshawks, woodpeckers, or old trees becomes precipitously high and recklessly controversial. The deferred opportunity cost potentially becomes too onerous, financially or politically, without these ubiquitous and incremental investments. Forest banks, carbon banks, mitigation banks, conservation easements, forest and agricultural preserve zoning, alternative taxation plans, urban tree ordinances, the conservation reserve program, the forest legacy program, best management practices, and best-value contracting are examples of incremental economic incentives for helping different kinds of forest landowners working with different forest types to maintain broad landscape integrity and a sense of place.

Although many of these strategies are beginning to be used, their long-term success and widespread adoption are constrained by the larger economic and political structures in which they are embedded. The fallacies of the market bedevil the economics of community forestry. For example, front-end capital investment strategies and back-end value-added niche marketing strategies are needed if transitions to a sustainable forest stewardship economy are to occur. The tail of niche market developments has failed to wag the dog of obtaining full investment values and returns for the full range of forest attributes. Niche market development, even green-certified niche market development, has more often than not become hostage to various kinds of national and international forest policy battles such as the "more cut or no cut" or "let burn or no burn" forest management debates. Lack of agreement on these bigger issues has severely limited the reliability and supply of even the low-value raw materials that value-added enterprises have needed to attract investment and stay afloat. Also, low-value forest residues, even with the most creative marketing and value-added schemes, can barely pay even the costs of deferred forest maintenance for the forest roads or the excessive fuel loads that are most directly associated with value-added market development. Saddling value-added and niche market microenterprises with the cumulative social, environmental, and economic legacy of chronic forest disinvestment is both inequitable and unworkable. Chronic disinvestment in the poorest urban and rural communities is associated with significant externalities that are beyond the purview and institutional role of niche market development. Sprawl, suburbanization, gentrification, and fragmentation of forest landscapes and forest communities are examples of market externalities that cannot be addressed by investing in value-added and niche market product development alone.

Economic policies need to encompass asset development and other questions in order to begin addressing the engagement, exchange, and investment needs that community forestry practitioners and supporters have identified. How are community engagement and community–forest exchange issues addressed so that forests are managed sustainably for their full range of attributes and benefits? Second, what mechanisms for community engagement and exchange are most equitable and feasible at the relevant ecological and economic scales, and how is that determined? At any given time and in any given place, the appropriate economic, social, and environmental scales for investment analysis and portfolio development may be different and may evolve differently. In particular, investment discussions need to explicitly address the effects of market imperfections on forest investment strategies, and they need to explicitly identify and strengthen the links between forest investment and community well-being. The next section presents a way of thinking about the connections between investment, equity, and social justice as they relate to community forestry.

Forest and Community Asset Building

Although investment strategies are beginning to increase the flow of value back to forest ecosystems, albeit in preliminary and experimental ways, these strategies and the revenue streams they produce rarely benefit the poorest segments of rural or urban communities. Social equity and justice concerns often are left unaddressed. Many investment strategies operate at either the forest landowner or regional scale; few are community scale. Thus a contradiction arises: Community forestry is founded on the presumed interdependency of forest ecosystem health and community well-being, yet the primary modes of investment and reinvestment to date have focused more on personal income rather than on the interdependence of community and ecological investment needs. For example, most approaches to poverty alleviation are still rooted in trickle-down economic theory or individual income and family-level poverty alleviation. These policies ignore the interrelatedness of community and ecological disinvestment and impoverishment.

The capitals framework, the asset-building approach, and the concept of ecological poverty highlight the relationships between environmental degradation and community impoverishment. They focus attention on the relationship between degraded environments, the distribution of resource access and use rights, and chronic poverty. Using the capitals framework as a conceptual underpinning for forest management and for poverty alleviation policies has significant implications. The capitals framework emphasizes the development of investment strategies designed to address forest ecosystem health and community well-being rather than relying solely on trickle-down economic policies and personal income enhancement social programs. The capitals framework explicitly identifies the various potential dimensions along which investment may occur, it facilitates the process of determining how to most effectively develop interlinked and integrated forms of investment, and it provides the rationale for why investment is needed across multiple assets, natural and social. It also illustrates the importance of developing tight institutional and scientific links between investments in natural and community capital. In short, the capitals framework implies that to be effective, social and economic policies must address community impoverishment and environmental degradation in an integrated fashion, often at the community scale. This entails ensuring that community-scale solutions are included in the suite of possible policy alternatives.

A primary benefit of the natural and community capitals framework is its ability to identify and prioritize the multiple dimensions and forms of investment. This is especially relevant where chronic community impoverishment is co-located with cumulative environmental degradation. For example, whether or not chronic disinvestment in natural and community capital in inner cities is partly the result of overreliance on short-term, individually focused, "upward mobility" policies has significant implications for the design of effective poverty alleviation strategies.[5] In places characterized by both chronic environmental degradation and socioeconomic decline, strategies that focus not only on individual income improvement but also on communitywide asset improvements are fruitful and should be promoted, implemented, and evaluated. Considering the sustainability and asset maintenance of natural, social, human, physical, and financial capital as integrated goals enables the development of integrated solutions to related problems. Integrated solutions are relevant, right, and just at all scales from local to international and from short term to long term. Conversely, the interest-balancing model operates best at the largest scales because it relies primarily on short-term mitigations for addressing environmental, economic, and social costs. Mitigations are largely ineffective at the smaller spatial scales and the longer temporal scales that characterize community forestry goals and activities.

Some public leaders and elected officials have recognized the importance of strategic, community-scale investment policies to correct the unequal effects of capitalism and to ensure overall political and social stability.[6] Public and philanthropic leaders whose primary allegiances lie with impoverished communities advocate strengthening the capacity of communities that the market has marginalized to attract the multiple forms of capital investment such communities need.[7] Without adequate financial and political resources, these groups remain particularly vulnerable to the negative effects of various forms of environmental degradation such as deforestation, desertification, and exposure to toxic wastes. Many of these groups, both urban and rural, now articulate their concerns and issues in terms of environmental justice. Having experienced the interrelated and oppressive effects of impoverishment and environmental degradation, urban and rural communities, particularly communities of color, are working to build their capacity to fight for clean and productive environments, high-quality job opportunities, affordable housing, and related issues.

There is clearly an unmet need for forms of investment, analysis, and policy that simultaneously address the interconnected problems of impoverished communities and environmental degradation. Conceptually, the capitals framework has been advanced to emphasize the analytical potential of strengthening connections between economic policy, environmental health, and community well-being. The following case study illustrates some of the possible links between investment opportunities and community-scale solutions.

Valuing a Fuller Range of Forest Attributes: A Case Study

Relations between the rural headwaters of the Feather River in the Sierra Nevada and the urban centers of southern California, destination for much of the Feather River's water, illustrate the challenges and the rewards of building and linking community capacity and well-being with forest and water conditions in both urban and rural settings.

The collaborative, community-scale watershed restoration efforts in the forested headwaters of the Feather River in northern California, a linchpin water source for the California State Water Project, are groundbreaking for many reasons. They illustrate the importance of moving beyond balancing competing interests to integrating them through capital formation at the community level to rejuvenate exhausted and overallocated environmental resources. They show the possibility of designing investment strategies that expand natural capital and build community capacity, and they illustrate the ripple effects that increasing social capital in one arena can have in other settings. In southern California metropolitan regions, destination for much of the river's water, urban community-based groups are developing innovative, grassroots-based programs that integrate ecological and urban renewal. Incipient links and common interests between urban and rural communities are emerging. One physical connection between these rural and urban areas stems from the potential reduced reliance on upstream water supplies that results from investing in strategies to restore urban watershed function, partially through better forest management in both urban and rural settings. Community-scale investment strategies in Los Angeles create options and opportunities not only in southern California but also in the upstream rural communities. Efforts are being made to institutionalize these relationships at a policy level through the nonpoint pollution sections of the Clean Water Act and through building urban–rural community coalitions around environmental justice policy development and implementation in water and energy development and in land use planning.

Because of the immense ecological, financial, and social values of water originating from the forest ecosystems in California and the high-stakes politics associated with efforts to preserve or concentrate these values, detailed studies are available that estimate and document the flows of value and regional disinvestment trends. The numbers are sobering, but the opportunities are encouraging. In terms of disinvestment, the trends are unmistakable. The combination of the "natural" (watershed) and manufactured (engineered) infrastructure enables water and hydropower purveyors to produce significant wealth through water and power sales. Most water supply and hydropower purveyors are accustomed to investing a portion of their profits from water sales into maintaining the ditches, dams, turbines, and pipelines that harness the watershed's runoff. Most water purveyors are unaccustomed to making similar investments in maintaining the watersheds that actually collect the "favorable flows" that are harnessed for economic uses downstream. This is especially true when the watersheds above the water supply or treatment facilities are not owned by the water purveyors themselves. The result is long-term disinvestment, the magnitude of which was revealed by the Sierra Nevada Ecosystem Project and other ongoing studies.

The congressionally authorized and funded Sierra Nevada Ecosystem Project determined that water accounts for more than 60 percent of the $2.2 billion worth of commodities and services the Sierra Nevada ecosystem produces annually. Stewart (1996:1019) estimates that the right to divert water from the Sierra Nevada is worth approximately $1.5 billion annually. Municipal water supply and hydropower interests reap more than $1.32 billion annually in profits from favorable flows from Sierra headwaters. However, they pay essentially nothing for watershed maintenance. In 1994, annual hydroelectric revenues deriving from the Sierra Nevada rivers totaled approximately $610 million, and downstream municipal and agricultural sales of Sierra water totaled approximately $800 million (Stewart 1996:1018).

A large proportion of these values derive from the Feather River. The Feather River originates in the rural Sierra headwaters above the California

State Water Project's uppermost dam, Lake Oroville, which provides 25 percent of southern California's state water contractor's supply for 20 million people. In 1996, the total value of Feather River water was approximately $427 million, of which $119 million was for downstream hydroelectric customers, $257 million for downstream municipal water users, and $51 million for downstream agricultural water users. This is approximately 31 percent of total economic value of waters originating from the entire Sierra Nevada range in 1997 (Tom Hunter, personal communication, 1999).

Given the large flows of profits from the sale of water and hydroelectric power generation, proposals for assessing water user fees have been developed in an attempt to generate revenue to reinvest in watersheds and communities. The revenue-generating potential of water user fees is significant. For example, a 1992 National Conference of State Legislatures study calculated that a $50 per acre-foot water user fee, if applied to the 39.3 million acre-feet of water annually used in California, would raise more than $1.9 billion in annual revenues (Allen 1992). Modest water user fees can equal or exceed the value of other resource commodities. A 1994 study by the University of California Wildland Resources Center suggests that on national forests in the Sierra Nevada, the revenue generated by water user fees could more than make up for the loss of receipts associated with declining timber harvests. Approximately 19 million acre-feet of the state's water originates from national forests. A $10 per acre-foot user fee (equivalent to 0.0003 cents per gallon) would generate $190 million, which is significantly greater than the average annual stumpage value of $154 million for harvests from public forests during the 10-year period from 1984 to 1993 (Stewart 1996:1023).[8] Although water user fees have yet to be implemented, they could be most easily collected directly from water and hydroelectric purveyors, who would pass the fees on to their customers. Fees could be collected by the State Board of Equalization, the State Water Resources Control Board, or the California Resources Agency. They would then be allocated to local, state, or federal (such as the Forest Service) entities for community-scale ecosystem reinvestments, possibly through forums similar to the Resource Advisory Councils established by counties to oversee implementation of federal funds provided through the County Payments Bill.

The investments in forest and watershed restoration and maintenance that water user fees could enable would enhance water supplies and, maybe more importantly, prevent further deterioration of water quality. A growing body of literature links worldwide deforestation to declining water supplies and worsening water quality. At the same time, policy makers and other innovators are increasingly realizing that the hydrologic services forests provide are among their most valuable and that investments in ecosystem management often are more cost effective than investments that increase water supply or treat polluted water (Johnson et al. 2001). The tradeoff between ecosystem management investments that prevent further water quality deterioration and after-the-fact water treatment is particularly evident in California. Currently, in California declining water quality has triggered the development of costly Total Maximum Daily [pollution] Loads (TMDLs) for 144 water pollutants under the nonpoint source sections of the Clean Water Act. A total of 129 water bodies (lakes, rivers, bays, and estuaries) are listed as impaired by the State Water Resources Control Board and the Environmental Protection Agency.[9] Developing TMDLs costs the state $11.5 million annually. The state needs to develop another 400 to 800 TMDLs for 1,500 listed impairments to comply with the Clean Water Act, without any new pollutant and water body listings. Developing a TMDL costs from $300,000 to $500,000 to adequately evaluate, allocate, and enforce pollutant cleanup responsibilities among the polluters in the watershed of an impaired water body (Thomas Mumley, personal communication, 2002).

Recognizing the expense and difficulty of after-the-fact pollution control efforts, let alone the overwhelming number of TMDLs that would have to be developed, the Public Advisory Group established by state law AB 982 (1999) to help guide the development of the state's TMDL program has called on the State Water Resources Control Board to "encourage the development of initiatives that focus on watershed-based approaches to attaining water quality on a holistic basis, and not merely elevating TMDL development above all else." Furthermore, they suggest that "the State and Regional Boards should return their primary focus to watershed management and realize that TMDLs are a tool to achieve water quality goals within the watershed context" (Caucus of Regulated Representatives 2001:2). Viewed in this context, water user fees can become an important tool for raising the capital needed for the sorts of incremental, watershed-based approaches for which groups such as the TMDL Public Advisory Group are calling. The state's biggest water users now have more of a vested economic interest in cleaning up their water supply sources. They can invest incrementally and cooperatively in monitoring and enhancing overall watershed health before a TMDL is developed. Once the State Water Resources Control Board or the Environmental Protection Agency orders a TMDL, the compliance options become more limited and more costly for the largest polluters. The political feasibility of user fees for source water quality protection is enhanced when science-based, partnership-based, watershed-based, and region-specific approaches are used and when the transparent intent of collaborative efforts is to comply with pollution and safety standards rather than evade or weaken them.

Watershed rehabilitation efforts by the Feather River Coordinated Resource Management group, a local watershed group, exemplify community-scale investment strategies that improve and protect water quality and, by protecting water quality, reduce the need to develop, implement, enforce, and monitor extremely costly TMDLs. The Feather River Coordinated Resource Management group has demonstrated a new "geomorphic" approach to fixing eroding and gullied stream channels and dewatered meadows. Preliminary results indicate that successfully reconnecting the meandering channel with its naturally evolved floodplain has extended the period of stream flow from ephemeral to perennial flow, moderated the magnitude and duration of peak flow events, reduced seasonal groundwater fluctuations, and prolonged summer base flows in the project area and downstream in Last Chance Creek (Lindquist and Wilcox, 1999).

This watershed restoration work has brought significant local economic benefit. Since 1990, the Feather River Coordinated Resource Management group's watershed rehabilitation projects have injected nearly $10 million into the local area through the wages and salaries of contractors, their employees, local consultants, and the group's staff. These projects provide additional opportunities for grading contractors and their operators to supplement their normal road-building, land development, and utility installation activities with contracts that use their skilled operator training and experience. This effort has coincided with diminished road building associated with the reduction in timber harvesting regionwide. Over the past 11 years, this has translated into 110 industry standard jobs. These jobs have typically spanned 1 to 3 months, with operator wages compensated at State Prevailing Wage or federally required (Davis–Bacon) wage rates. To implement these projects, studies required by the California Environmental Quality Act and National Environmental Policy Act have been conducted. These activities have provided numerous consulting contracts for local professional biologists, botanists, hydrologists, engineers, and archaeologists to work in their fields and in their community. The Feather River Coordinated Resource Management program alone has supported a two- to three-person staff full-time for more than a decade to coordinate and implement this program. When contrasted with the alternative solutions to upland erosion, such as costly dredging and silt removal from downstream reservoirs, it is clear that upland watershed restoration and erosion prevention are a more effective and socially desirable way to restore watershed function and invest in community capital.

Community-scale restoration efforts in the forested headwaters of the Feather River are complemented by analogous efforts in downstream metropolitan areas of southern California to revitalize both communities and their watershed. These once disconnected efforts are increasingly hydrologically, socially, and politically linked with community-based upland forest and watershed restoration efforts such as the Feather River Coordinated Resource

Management efforts. As with rural hinterland areas, the benefits of investment strategies that replenish depleted natural and community capital in urban settings are manifold. For example, ultra-urban areas (areas with more than 80 percent impervious surfaces) produce excessive runoff, which destabilizes and pollutes receiving streams, lakes, and bays. Reforesting the inner city can simultaneously reduce this runoff and provide desperately needed green urban spaces. Replacing cement with greenery redirects storm runoff into stormwater infiltration and groundwater recharge. In this manner, supply augmentation could conceivably result in less water diversion from upstream headwater regions.

Strategies for regreening the inner city for runoff retention and energy conservation can include investing in community-scale activities such as urban forestry and reforestation, urban gardens, and the kinds of brownfield redevelopment that increase urban forests and wetlands. Urban water is also recharged, detained, and cleansed by capturing stormwater runoff in other ways. These include restoring urban creeks and re-creating riparian forests on urban floodplains, installing permeable rather than impervious paving, and developing vegetated floodwater detention areas for runoff from impervious and polluted roadway and parking lot surfaces. Investment activities such as these help transform stormwater into a community and environmentally friendly asset. These and other strategies for restoring watershed function in ultra-urban areas strengthen the interdependence between community and landscape and, though indirect, offer unique and potentially powerful ways to link urban with rural residents and riparian systems with forest health.

Urban community forestry and watershed groups are just as attuned to the relationships between building community and natural capital as their rural counterparts. Some have begun to quantify the benefits of community-scale solutions to urban water issues as one way to advance their cause. One example of this is the Los Angeles and San Gabriel Rivers Watershed Council, which focuses on ways local water harvesting can reduce dependence on water imports. In one of the Council's publications, authors Dallman and Piechota note the irony of simultaneous investments in both urban stormwater drainage and long-distance water imports and identify alternative investment strategies to damming and diverting more water from the Colorado River and northern California rivers (1999:40). The report's authors calculate that if 80 percent of rainfall were captured from only 15 percent of the total watershed, then total runoff would be reduced by 30 percent. This would be the equivalent of 132,000 acre-feet (or enough to supply 800,000 people for a year), which would not have to be diverted from the Colorado River or northern California rivers.

In Los Angeles, TreePeople, a 25-year-old urban forestry group, has begun quantifying the benefits of designing urban landscapes to function as forested miniwatersheds. Preliminary results suggest that if the watershed approaches they outline were widely implemented in Los Angeles, then freshwater imports to the region could be reduced by 50 percent, pollution into Santa Monica and San Pedro Bays would be dramatically reduced, the 100-year flood threat to Los Angeles would be eliminated, 30 percent of the city's landfill capacity would be made available through green waste recycling, and air quality and energy efficiency would be improved in buildings and through a general reduction in the heat island effect (Condon and Moriarty 1999:10).

Reforesting the inner city for water and energy benefits opens new institutional pathways for actualizing environmental justice in a more proactive and participatory way. Empowering local communities to participate in the design and implementation of decentralized, community-scale redevelopment—green infrastructure developments—creates the opportunity to build community and natural capital simultaneously. These efforts also begin to make explicit the linkages between urban and rural areas, in terms of both conceptual analogies and shared key issues, and common ecological relationships based on the "forested watershed" concept, which links downstream users with upstream sources of water and energy.

By piloting a redirection of a portion of existing public water, wastewater, and flood control infrastructure budgets to natural asset restoration and maintenance in and around impoverished communities, these cases confront conventional mitigation and trickle down policies for addressing community impoverishment and environmental damage. The degradation of forests and watersheds is most evident in proximity to the poorest communities in both inner-city and remote rural areas. Lack of investment by urban water users in maintaining the ecological functions and the species diversity of urban and rural forests is increasing social, environmental, and social conflicts, with disproportionate impacts on the poorest communities. Investment strategies, such as those in the Feather River's rural headwaters and throughout its urban service area, are needed to both create and redirect revenues for forest and watershed restoration approaches and to rebuild well-being in resource-poor communities.

Local Responses to Barriers to Equitable Engagement and Exchange

Community activists from urban and rural areas are coalescing around an increasingly conscious and collaborative strategy for alleviating both environmental degradation and community impoverishment. By questioning the prevailing bias toward big solutions to big problems, community groups such as those described in this chapter are challenging the pervasive science and investment subsidies for large-scale, one-size-fits-all, top-down solutions to our nationwide forest, energy, and water problems. Community activists are wrestling with the contradictions that plague national-scale natural resource problems and policies. On one hand, community-based initiatives need to address site-specific complexities; on the other hand, government administrative policies and procedures need to be consistently applied across millions of acres and for millions of Americans. Confusing consistency, complexity, and equity in investment and mitigation policies is a significant barrier to community-scale capital formation.

Community-scale solutions differ from large-scale solutions in substance and process as well as in size; they are not simply smaller models of large-scale fixes. Often, community-scale solutions entail collaborative processes that involve broader definitions of problems and stakeholders. Community groups are beginning to assert that single-purpose, single-interest solutions often do not include the synergy of truly integrated decision making even if multiple stakeholders are involved. Interest-based processes tend to discourage thinking outside the box. By their very nature, they often limit the problem definition to one or a few interests and thereby exclude integrated solutions to linked problems. With overly narrow definitions of problems and of stakeholders, traditional interest-based "balancing" and conflict resolution processes often exacerbate the decline of ecological and cultural assets over time. Interest-based decision making is generally top-down rather than a melding of top-down, national-scale goals and standards with bottom-up, site-specific community-scale approaches for meeting those goals and standards. When national-scale goals and standards can be melded with community-scale approaches and alternatives, the balancing broadens. At local and larger scales, the broader goal becomes balancing short-term effects with long-term ecological, social, and economic asset conservation.

Community-scale endeavors often entail more sophisticated (finer-grained and smaller-scale) social and ecological analyses and nuanced forms of investment than the blunt investments associated with large-scale wood, energy, and water product development. Because small solutions often are invisible, at larger scales they are perceived as insignificant contributors to solving large-scale problems. Equitable access by communities to monitoring, modeling, and state-of-the-art science at the community scale is a substantial barrier to community-scale solutions receiving equitable financial and technical support. This is especially true for impoverished communities.

Using the following strategies, some community groups have begun to transform the stranglehold that trickle-down and single-interest solutions have on urban and rural, environmental, social and economic policy, and investment practices. These strategies include

Utilizing participatory science to inform and legitimize bottom-up solutions to national-scale problems.

Building community dialogues and networks that bridge the urban–rural divide in forest management and poverty policy through more inclusive policy development and more community-sensitive impact assessment processes.

Empowering community coalitions seeking to meet or exceed regulatory standards through community-based solutions to become equal participants in policy- and decision-making venues and to have more equitable access to technical and financial resources.

Where political decision making is polarized to the point of gridlock or wholly captured by narrow interests, the following strategies have been effective for some of the community groups who are seeking to elevate community-scale solutions to national problems:

Allying with environmental, corporate, or government interests as junior partners rather than being marginalized as third-party impacts to be mitigated.

Establishing themselves as interest group players in their own right by brokering new relationships between traditional environmental, industry, or government stakeholders.

Establishing special relationships with powerful outsiders such as property rights coalitions, faith groups, labor, key legislators, key regulators, public trust advocates, environmental justice advocates, social justice advocates, and independent scientists.

Wedging broader public involvement into traditional insider decision-making processes by requiring voter ratification of decisions or by restructuring legislative reauthorization, bonding, or funding authorization processes to include minority reports, third-party monitoring or certification, independent science review, and grant making for community-scale pilot project testing.

Conclusions

Ensuring social, environmental, and economic justice at all scales entails developing multiscale asset conservation strategies and securing effective access to decision-making arenas and the equitable distribution of financial and technical resources. Endangered Species Act enforcement actions, skyrocketing local joblessness and local business bankruptcies, combined with media-worthy civic breakdown at the community level are the all-too-common failed report cards of an overly narrow reliance on interest-based balancing for resolving both long-term asset conservation and short-term equity questions. Ironically, it is the community level, closest to the ground, that is the earliest warning system for identifying when the interest-balancing trickle-down costs and benefits system is dangerously off center. Distant interest groups and the larger public, in contrast to the local communities, may have no interest in early resolution of conflicts in overallocated, undercapitalized systems. At the crisis moment, the nation may take notice of conflicts, such as the standoff between local farmers, environmentalists, American Indians, and government agents in the Klamath Basin in northern California. But by then it is often too late for the affected communities, human and natural. Early negotiations could result in equitable integration if the whole "competing needs" process is not allowed to push the overall demand curve ever upward and ever further away from communities and their local landscapes.

The potential value of community-scale solutions lies in their inherent flexibility and their ability to track and remediate cumulative effects. Community scale tends to be associated with small-scale and incremental solutions to a set of linked problems rather than problem solving in ways that overcommit resources to more irreversible, less adaptable, and narrower courses of action. Overscaling solutions and overcommitting resources prevent adaptive management and learning and the midcourse corrections that are the essence of ongoing sustainability. Too late, too single-purpose, and too big solutions disproportionately and inequitably foreclose options at the (ecological and human) community scale. Options are replaced with mitigations. Mitigations, such as retraining for laid-off workers, provide individuals or families with more options, but mitigations rarely focus on rebuilding social and ecological assets at the community scale and over the necessary reinvestment time frames for effective capital formation.

Scientific indicators and assessments for environmental and social capital and for environmental and social justice are unavailable, undeveloped, or not integrated at the necessary scales of analysis. If the assessments of the function, condition, and trends of assets are unavailable, then the quantitative basis for investment is also lacking. Integrated investments take integrated assessments. Analyses of resource use efficiency and resource use equity are not substitutable for one another and should not be confused, as they often are. Large-scale snapshot assessments and community-scale, site-specific, multigenerational knowledge are not interchangeable kinds of information or analyses. They should not be confused, yet they usually are. Without adequate analysis, there will never be adequate investment.

Another major difficulty for equitable treatment of community-scale solutions is that a much greater front-end investment in social and human capital is needed. Integrating top-down goals and standards with community-scale, site-specific solutions and incorporating environmental and social justice into asset formation and conservation policies and programs entail preinvestments in enhancing community capacity. Some philanthropic foundations have been funding part of the requisite multicultural, multi-issue community capacity building that community-scale solutions often need. These foundations recognize the limitations of public involvement processes when (as is often the case) community groups lack the technical and financial resources to participate effectively in such processes. This is especially true in multicultural and multilingual or transient community contexts. Beverly Brown of the Jefferson Center in Oregon, who was interviewed for this study, describes the challenges of working with diverse multicultural and multilingual groups of workers in the following manner:

> We have cooperated with the low income and mobile multicultural and multi-lingual contingent labor force for six years. We have learned that in spite of the daunting challenges of contingent labor, people make it work because they invest in strong social networks. Each different language and culture adds a factor of three to five to the complexity of building community capacity compared to middle-class, white-only and single-occupation labor organizing.

At the back end of the process, there is inadequate attention to documenting and extrapolating from isolated experiments. The potential of a thousand small solutions versus one big one therefore is often underestimated. When one moves to the full attributes of forests and full equity issues at community and larger scales, the range of alternatives also becomes larger. For big projects, the design and feasibility analyses and the operation and maintenance costs are nuts-and-bolts kinds of engineering costs, which are often supported with public funds. For incremental solutions using small-scale projects, the preproject investment usually is in building community capacity for design and implementation and for evaluation and adaptive learning, which are rarely funded by public dollars.[10] Community groups often are unaware of their options soon enough to prevent "win–lose" crises and conflicts. Science and public education are only moderately successful at compiling and disseminating information about community-scale management and technology successes in culturally appropriate ways. Often community representatives take on a circuit-riding role to make communities aware of opportunities, represent community voices at policy tables, and bring technical assistance to communities. Here's how one circuit rider describes the process:

> Circuit riding is about consistent faces in many places. By having a consistent presence in policy-making venues, artificial divisions can-

> not be drawn between stakeholders and communities. Policy makers begin to expect community perspectives and adopt community language. It's about power sharing. Community-based solutions threaten existing power dynamics, by their very nature. They are site specific, incremental, flexible, and use innovative, small-scale technology. They are partnership-driven, decentralized solutions and they decentralize, rather than consolidate power. Because as circuit riders, we are linked at the "problem shed" scales, we become consistent, dependable voices that break up the stereotypes that keep communities divided, isolated, and disempowered. (Lynn Barris, personal communication)

Discussions of investment revolve around the key themes of this book. When viewed from the perspective of the people on the ground, of practitioners and activists, the multiple challenges that community forestry represents for science, politics, and economics become clear. What also becomes evident is the promise that community forestry holds for developing community-scale solutions to the pervasive problems of environmental degradation, community decline, and poverty. These solutions are effective at multiple scales, and they are finding increasing acceptance in nonlocal decision-making arenas. When they are informed by environmental justice concerns, they empower politically and economically marginalized communities. They are also forging a new model for the practice of science, one based on participatory processes and decentralized mechanisms for adaptive feedback and monitoring. Finally, by insisting that the linkages between natural and social capital be identified, explored, and used as the basis for developing investment strategies, community-scale community forestry remains true to its claim that its purpose is the twin goal of improving community well-being and forest ecosystem health.

CHAPTER 10

Signposts for the Twenty-First Century

> . . . environmental policy in the United States is in the early stages of what could be a profound transition toward sustainable communities.
>
> —*Mazmanian and Kraft (1999:285)*

Historical Continuities

A gaze backward in time shows that community forestry at the beginning of the twenty-first century shares much with the community-focused and socially responsive forestry traditions that existed at the beginning of the twentieth century. The practices associated with the diverse community forestry traditions of a century ago along with the ideas of Progressive Era thinkers such as Benton MacKaye resonate strongly with the activities and the ideology of the current community forestry movement. The similarities are striking.

Hispano water and forest management traditions in the Southwest are one of the most palpable examples of the links between community, collective action, and shared resource dependence, all central themes for the current community forestry movement. Seen especially within the social organization of *acequias,* traditions of community-based collective action in Hispano communities reinforce and strengthen the meaning of community. In community forestry, shared dependence on forests constitutes the basis for developing collaborative arrangements for forest management that simultaneously build community. This points to the synergistic relationship between community well-being and forests that lies at the heart of community forestry. Among Hispano communities the relationships between community, cultural identity and continuity, and resource management are particularly strong.

Maintaining access to resources for groups that depend on them for their well-being is another important theme that emerges from discussion of Hispano resource management. The gradual undermining of Hispano access to forestlands they had used for generations created resentment and distrust, contributed to the depressed economic condition of the region, and made it difficult if not impossible for dialogue and collaborative resource management efforts to take root. The importance of access to forests and forest resources for current community forestry groups is a continuing and central concern across the country. Like the discussion of collaborative arrangements between public lands managers and indigenous groups, the central importance of access for Hispano communities underscores the need for collaborative resource management on public lands that includes communities with traditional and customary rights as well as other stakeholders.

Place-based local knowledge and traditional ecological knowledge, embedded in customary practices and encoded onto forest landscapes, were hallmarks of prior community forestry traditions. They are also central elements of the current community forestry movement. It is becoming increasingly apparent that Native American groups, rural communities, and workers possess fine-grained, site-specific knowledge relevant to ecosystem restoration and often are well positioned to monitor and evaluate forest ecosystem changes over time. Because of the tight coupling between knowledge and power, diversifying the forms of knowledge accorded legitimacy in resource management professions entails sharing power with people and communities who have local and traditional knowledge. It involves the devolution of decision-making authority to more local levels by land management agencies because of the site-specific and local nature of these forms of forest knowledge. It also involves a willingness by scientists to share power and create opportunities for genuine two-way exchanges of information. This, in turn, entails acknowledging the legitimacy of the knowledge of people intimate with the land and their perspectives on forest management and ensuring adequate levels of access to forest management processes and to the resource itself, both of which are key community forestry themes. Similarly, recognizing the validity and legitimacy of the traditional ecological knowledge of indigenous groups relates directly to the argument that workers and community residents that live near, use, and work in public and private forests have developed understandings and perspectives that are important for forest resource management, stewardship, and restoration. Community forestry also involves developing rigorous social and ecological monitoring and evaluation protocols. Developing and using monitoring procedures, especially collaborative all-party or multiparty monitoring processes, is central to the feedback and learning processes associated with a more responsive science, adaptive management, and collaborative resource management.

The tradition of community forestry in New England firmly links current community-based forestry to urban areas and urban concerns. Community forestry in New England arose out of settlements, not from sparsely inhabited rural hinterlands. Given its roots in community and settled landscapes, the legacy of community forestry in New England provides opportunities for maintaining and enhancing green pockets in a context of land fragmentation, for using urban community forests as vehicles for strengthening civic culture, and for making connections between urban and rural communities and issues. It also speaks to the potential for community forestry to identify investment mechanisms and generate revenues that can achieve collective goals and lead to reinvestment in the ecosystem itself. This includes providing assistance to private woodlot owners in terms of technical expertise for sustainable forest management and product marketing and the creation of loan programs specifically developed to meet the needs of forestland owners. Institutionally, the visions of early foresters such as Bernard Fernow and Samuel T. Dana provide templates for rethinking the nature of the relationship between the federal government and urban forests and communities.

Benton MacKaye's articulate calls for the federal government to concern itself with labor issues in the national forests and for developing forest management techniques that further community capacity and civic culture resonate loudly with the articulation of forest worker issues in the community-based forestry movement. Echoes of MacKaye's arguments can be heard in current attempts to rethink the relationship of the Forest Service with communities adjacent to national forests and with the current workforce through mechanisms such as stewardship contracting. Additionally, MacKaye's trenchant critique of the ecologically and socially maladaptive workings of the market economy adds historical depth to current analyses of natural and social disinvestment and their cumulative damaging ecological and social effects.

Historical Discontinuities and the Extent of the Challenge

Despite the strong parallels between the current community forestry movement and its historical antecedents, the differences between them are also pronounced. The current community forestry movement harbors no romantic vision of a return to a halcyon past. Indeed, in almost all respects the barriers and challenges that the community forestry movement faces are far more daunting than those that existed a century earlier. This is because the intervening decades have allowed ample opportunity for the development and institutionalization of policies, practices, political alignments, economies, and interests that are inimical to community forestry. In a nutshell, many elements of the dominant model of mainstream, traditional, Progressive Era forestry represent the precise barriers and challenges that the community forestry movement must work to overcome if it is to succeed.

The challenges community forestry faces are significant and in many respects quite daunting. This is because community forestry represents nothing less than a radical realignment of relations between people and forest ecosystems. Community forestry is driven by a vision of forestry in which the sustainability of the communities and workers that depend on and steward forests is part and parcel of the long-term ecological sustainability of the forest ecosystem itself. Therefore, community forestry forges different relationships between communities and workers and forests. Community forestry challenges the inequity and social injustice that have historically accompanied forest resource extraction, and it makes the claim that achieving ecological sustainability is not possible without also achieving social sustainability. In this vein Raymond Williams (1980:85) notes,

> Out of the ways in which we have interacted with the physical world we have made not only human nature and an altered natural order; we have also made societies. It is very significant that most of the terms we have used in this relationship—the conquest of nature, the domination of nature, the exploitation of nature—are derived from the real human practices: relations between men and men. . . . If we alienate the living processes of which we are a part, we end, though unequally, by alienating ourselves. We need different ideas because we need different relationships.

Community forestry advances the idea of different relations, both between people and between them and forests. By arguing that the ways in which we interact with nature reflect and constitute relations between people, Williams's arguments lead to the conclusion that not only are forests and communities interdependent, as the community forestry movement claims, but also that healthy forests are not possible without thriving communities. Thus, it is an oxymoron to attempt ecosystem management tasks such as thinning or fuels reduction without also attending to the labor relations and working conditions through which that work is accomplished. This is the challenge of community forestry: to forcefully and convincingly make the case that social sustainability and ecological sustainability are interdependent phenomena and then to effect the wide array of changes necessary for the simultaneous achievement of both.

The necessary changes are more daunting than they were a century ago. The effort involves unseating pervasive and entrenched institutional practices and relations within several different arenas. It requires that a model be advanced of socially just participatory democracy that avoids the exclusionary pitfalls of civic republicanism and substitutes multipolar community-scale integration for bipolar interest group–based policy-making processes. Achieving this necessitates attending to the challenging issues of how to enfranchise people and groups, especially workers and people of color, who have historically been disenfranchised from natural resource management decision-making processes. This process entails the rehabilitation of work and occupation as legitimate bases for forest enfranchisement, in addition to the other institutionalized forms of enfranchisement: those of citizenship and territorial control. Considering the history of political disenfranchisement of many of these groups, the weak legal framework for protecting worker rights, and the lack of basic legal assets among some groups of undocumented workers, the challenges of building a movement that includes the enfranchisement of workers are hard to underestimate. Yet, as we argued in preceding chapters, the price of not working to enfranchise these groups is the long-term success of the community forestry movement, for given the strength of the forces inimical to community forestry, only a solid coalition of all those who depend on the forest will be able to muster adequate countervailing force.

A second challenge for the community forestry movement is to effect the changes that are necessary within the structure and practice of government to support community forestry's vision of healthy forests and healthy communities. This entails achieving a wide array of institutional changes in public lands management agencies and forestry extension on private lands. Although risk takers in state and federal government have established important precedents for what can be accomplished through meaningful engagement with communities, workers, and other stakeholders, the institutionalization and widespread legitimization of the kinds of collaboration they have pioneered have yet to occur. Securing the necessary organization takes clear leadership direction at all levels, implementation of accountability measures that support collaboration, and modification of internal organizational incentive and reward systems to reflect the importance of such activities. Unless and until such changes take place, community forestry in many contexts will be vulnerable to the personal predilections and preferences of line officers and extension foresters. Enfranchised communities, workers, and other stakeholders will have little or no leverage to push for community forestry without the requisite institutional changes at county, state, and federal levels.

A third challenge concerns the practice of science. Much has been made in this book of the inconsistencies between the dominant model of scientific practice, with its Progressive Era roots, and the types of scientific practice needed for community forestry to succeed. We advance a civic science framework for conceiving of the role and purpose of science. Seen in this light, the hegemony of western science is muted by the recognition that traditional ecological knowledge and other forms of local knowledge are vital complements to traditional, Progressive Era science, especially when confronted by challenging issues of ecological complexity and uncertainty. Recognizing the limits of western scientific knowledge and methods allows for the growth of the humility necessary to advance a deliberative, participatory science and to take monitoring seriously. When combined with the requisite institutional changes, this creates the conditions necessary for the implementation of all-party or multiparty monitoring. Being the linchpin of an adaptive management process, monitoring becomes an important tool that furthers the deliberative conversation by generating the information needed for learning and feedback to occur. All-party or multiparty monitoring is also an important vehicle for developing the trust among diverse parties that is a prerequisite for successful collaboration.

The fourth major challenge concerns the variety of institutional changes necessary to generate the revenue streams needed for community-scale forest ecosystem management and restoration. Although the search for value-added processing opportunities and the economic utilization of the byproducts of forest restoration are important parts of this effort, they are only small parts. From the great reservoir of value that forests contain and produce, much larger revenue streams must be tapped and made available for ecosystem reinvestment. Accomplishing this entails developing the institutional exchange mechanisms that allow recognition and valuation of the full array of benefits forests provide. This is a necessary step in realizing and then redirecting a portion of the revenue those benefits generate back into the ecosystem in a manner that sustains its ability to continue providing them. However, such investments verge on being irrelevant for community forestry unless they occur through community-scale processes and are tightly linked with poverty alleviation and the enhancement of community well-being. Forest ecosystem investments must help foster healthy forests and healthy communities through the simultaneous enhancement of both natural capital and community capacity. Seen in this light, reinvestment becomes an important vehicle for the simultaneous achievement of social justice and forest restoration objectives in both rural and urban contexts.

The emphasis on community-scale solutions and investment strategies is important for a number of reasons. As we have argued throughout this book, community-scale solutions are more likely to advance integrative approaches for addressing ecosystem management challenges characterized by uncertainty and complexity. Rather than the polarized "winner takes all" strategies associated with interest group pluralism, community-scale solutions involve crafting strategies that are supported at some level by all the involved stakeholders. When combined with local processes that embrace the "egalitarian politics of difference" (Young 1990:157) and are therefore able to avoid the oppressive workings of the "public interest," the ideal of community, and other paradoxes of democracy, community-scale solutions enhance equity and social justice at the local level. Community-scale processes are also essential vehicles for designing and implementing participatory monitoring strategies. Their inherent flexibility is an important element of adaptive management approaches to the "wicked" ecological and social problems associated with ecosystem management. And they offer important advantages over the blunt, single-purpose, macrolevel solutions that inevitably undermine more viable and equitable community-based processes and solutions.

Signposts for the Future

Achieving the vision of community forestry—the reconfiguration of relations between people and the forest ecosystems on which they depend so that both will be sustained—entails a radical revisioning of how we as a society structure relations between people and forests. The success of the community forestry movement depends on a wide variety of transformations in politics, government, science, and economics. In each of these arenas resistance to change stems from a combination of institutional inertia and the interests of those who benefit from the status quo. The entrenched nature of these forces requires that the community forestry movement, if it is to succeed, mobilize a broad base of support and resources to countervail them. Part of this process includes realizing the depth and breadth of the challenges community forestry faces. The systemic changes in institutions, social structure, and the allocation of value that community forestry must achieve are extraordinary. Going it alone, it will not succeed over the long term.

At least two major implications follow from the realization of the hurdles the community forestry movement faces. They both concern strategies for mobilizing the critical mass of resources for accomplishing the changes needed. The first, smaller implication is the acknowledgment that a broadly inclusive movement, as discussed in Chapter 6, is a necessary but not sufficient condition. As one workshop participant stated in this regard, "diversity [in community forestry] is not just nice, it is necessary." For the many reasons discussed herein, unless and until work and occupation are fully institutionalized in the movement as a valid basis for forest enfranchisement, the movement's gains will be ephemeral, vulnerable to the shifting sands of national-level political processes, flighty capital, global trade agreements, and other dynamics over which communities and workers hold little sway. Therefore, at a minimum, community forestry's staying power depends on its ability to empower all people and communities that have a stake in forest stewardship.

The second, larger implication is the need to identify and create alignments with other groups, organizations, and movements whose success depends on similar changes in social structure and the allocation of value. The community forestry movement is not alone in its calls for integrated solutions to environmental and social decline, participatory and just democratic processes, meaningful partnerships with government, civic science, and increased investment in communities and ecosystems. On the contrary, these and similar agendas for social change are shared by a diverse array of groups, organizations, and sister movements. Together they constitute a powerful network of similar interests whose resources, if mobilized in a coordinated fashion, could enhance the probability that needed changes would occur.

Four examples illustrate this point. The first gets to the heart of the community forestry enterprise: the need to develop integrated solutions to the interrelated problems of declining forest health and declining community well-being in a manner that promotes equity and social justice. With only slight modifications, this is the same underlying issue that drives the environmental justice movement, which has hitherto focused primarily on communities of color in urban contexts contaminated by toxic contaminants. In many key respects, the social change agenda of the environmental justice movement runs parallel to that of community forestry. In a manner analogous to the community forestry movement, environmental justice advocates call for the reversal of historic disinvestment patterns, development of mechanisms for increased investment and community-based economic development, and the political enfranchisement of historically disenfranchised people and communities. The overlap between the two movements is obvious (see Mutz et al. 2002). The shared challenges and the similar forms of social change for which the respective movements are working suggest that there is ample basis for strategic alliance building of the sort that will strengthen both movements and increase the likelihood that the institutional changes each call for will actually occur.

The second relates to the emphasis on community-scale solutions, civic engagement, and equitable and just democratic processes within community forestry. There are a large number of other organizations and sister movements for whom these are central concerns. Sister movements include civic environmentalism and the sustainable communities movement within the field of planning. Innumerable citizen groups and organizations around the country also have emerged in the last two decades for whom issues of civic democracy are absolutely core. All of these entities share a common focus on the need to replace the individualistic anonymity of interest group pluralism and its associated dependence on government to safeguard the "public interest" with public forums in which community-scale, face-to-face deliberation can occur in an effort to hash out shared understandings of challenging issues and problems and develop community-based strategies and solutions to them. The sorts of community-strengthening participatory democratic processes these groups and movements foster are quite similar to those of community forestry when they succeed. Similarly, the challenges they pose in terms of the meaning and purpose of government are also quite common. So, for example, the institutional changes in public lands management agencies needed for community forestry on public lands to succeed are analogous to those needed in other agencies such as the Environmental Protection Agency, the Department of Housing and Urban Development, state, regional, and local planning departments, and even law enforcement agencies. Key themes such as the need for meaningful collaboration, devolution of decision-making authority to relevant publics, and the institutional changes needed to support such endeavors are core elements of civic democracy. Together they paint a compelling picture of sea changes in the meaning of citizenship and democratic participation in this country.

The third example concerns the widespread and growing lack of confidence in science. The Progressive Era model of expert-driven science that serves the public interest is not just a casualty of the community forestry movement. The public's lack of faith in science's ability to find solutions to society's pressing problems is widespread and can be observed in sectors as diverse as medicine and public health, air pollution and global warming, and fresh and saltwater fishery restoration. It is worth noting that many of these issues contain the same attributes of complexity and uncertainty as those related to community forestry. Many groups and organizations have developed proposals for more public oversight and involvement in the scientific endeavor, from the joint identification of key issues and questions to the collaborative development and execution of the research itself, the interpretation of the results, and the deliberative development of the ensuing policy recommendations. Specific methods have been developed, such as Participatory Rural Appraisal, participatory research and monitoring approaches, and other protocols concerning participatory research that together demonstrate evidence of a sea change in the relationships between science and society. Thus the calls of the community forestry movement for civic science are not a lone voice in the wilderness; rather, they are part of a widespread surge of interest in reforming the practice, methods, and ethics of science.

Finally, if community forestry succeeds at rehabilitating work and occupation as a basis for forest enfranchisement, if it succeeds in making the case that forest social relations are as important as forest ecological relations, then it will have established a solid basis for effective alliance building with labor unions and other organizations that focus on worker issues. To date, labor union organizations have been, at best, only marginally involved in the community forestry movement and allied efforts such as forest certification, despite what appear to be clear overlapping agendas and interests. This lack of coordination is partly a result of the historical exclusion of worker issues from forestry debates, which, despite Benton MacKaye's calls to the contrary, have been devoid of serious consideration of the "problem of the lumberjack."

In addition to building alliances with organized labor, prioritizing forest social relations also entails working more closely with the unorganized, contingent workforce. Although working with contingent forest workforce issues and workers is an essential challenge for the community forestry movement, as has been argued throughout this book, doing so involves tracing the shadow lines that connect different sectors of regional economies. Because contingent workers cycle between different types of work, often on a seasonal basis, understanding their issues and concerns entails broadening frames of reference to include other sectors such as agriculture and fisheries, two of the main sectors where contingent forest workers also work. Forest social relations quickly spill over into and are connected with the social relations of other segments of regional economies. Therefore, effectively addressing them will require an understanding of how they articulate with these other sectors. At the same time, by broadening the movement's scope in this manner, the ability to generate a critical mass or momentum for addressing issues of shared concern will increase.

The survival strategies of contingent forest workers contain at least two lessons for the community forestry movement. The first lesson is that community forestry is situated within a broad, predominantly rural setting that includes a variety of other sectors, such as agriculture, fisheries, and tourism, as well as dynamic social processes including outmigration from rural areas, changing population demographics, and shifting land ownership patterns (industrial forestland concentration in some areas and subdivision and urbanization pressures in others). For many workers, communities, and landowners, community forestry is one of a variety of risk-reducing strategies that people engage with to maintain their economic viability in a context of dynamic change. Seen in this light, community forestry becomes part of an integrated community-based economic development strategy. Rather than a myopic focus on only the specific relations between forests and people, community forestry takes on new strategic potential and resilience and acquires new partners and allies when it is viewed as one element of integrated community-based economic development. No longer primarily the domain of resource management professionals, community forestry can and should become an important tool in the development toolkit of planners, community development practitioners, and the agencies and foundations that support them. One advantage of taking this more macro perspective on the community forestry landscape is that investments in various of the constituent capitals of community capacity, whose primary purpose relates to community forestry, can have important ancillary benefits as communities harness their strengthened capacity in the pursuit of other economic development objectives.

The second lesson that contingent workers offer the community forestry movement is more strategic than analytical. Through networks of friendship and kin, contingent workers are able to collect and assess information from multiple sectors and large regions regarding employment opportunities and conditions, barriers, and other factors. Based on their assessment of relative costs and benefits of different choices, they are able to develop a diverse portfolio of risk-spreading strategies that, if successful, enable them to meet basic needs.

The lesson for the community forestry movement comes into focus when these worker strategies are viewed as a metaphor for developing the movement's strategic potential. As the position of a single contingent worker is vulnerable given the strength of the forces that work against him or her, so also is the community forestry movement vulnerable if it is conceived of as an isolated movement, especially given the array of challenges discussed in the previous section. However, when community forestry practitioners and supporters use networks of common interest to scan the broader political and social landscape in which they are positioned, an astonishing array of risk-reducing and success-enhancing opportunities come into focus. By identifying common links with organizations and movements that share, at least in part, community forestry's vision and social change agenda, the community forestry movement itself becomes part of a broader network that has emergent qualities and collective strengths and resources. In many respects, the emergent qualities of this network derive from shared critiques of the dominant ways of organizing politics, government, science, and economics in the United States. The network nodes, in addition to the community forestry movement, include the environmental justice movement, other community-based movements that embrace community-scale processes such as civic environmentalism and the sustainable communities movement, community-based economic development, organizations advancing participatory research and civic science, conservation organizations, labor unions, and organizations working to advance the interests of contingent workers. When viewed from this perspective, the objectives of the community forestry movement are seen for what they are: nested calls for social change that resonate with other transformative processes across the country. Much like the many "spontaneous ignitions" that the early forms of community forestry assumed in the western United States, the different nodes of this emerging network have also arisen within specific contexts that give them their distinctive forms, structures, and practices. And much as the spontaneous ignitions of the western United States and elsewhere across the country gradually coalesced into a cohesive call for restructuring relations among people and between them and forests, so is the emerging network capable of articulating common ground and advancing a cohesive social change agenda. The likelihood that the individual nodes of the network, including community forestry, will be able to achieve their respective objectives depends, to a significant degree, on the strength and cohesiveness of the whole network.

APPENDIX

Study Interviewees and Workshop Participants

People Interviewed[1]

Bridget Abernathy, Mountain Association for Community Economic Development, Berea, KY

Monica Armster, National Network of Forest Practitioners, Tallahassee, FL

Adela Backiel, U.S. Department of Agriculture, Washington, DC

Conner Bailey, Auburn University, Auburn, AL

Taylor Barnhill, Southern Appalachia Forest Coalition, Asheville, NC

Jim Beeman, Hiawatha Sustainable Woods Cooperative, Fountain City, WI

Louis Blumberg, California Department of Forestry and Fire Protection, Sacramento, CA

Beverly Brown, Jefferson Center for Education and Research, Wolf Creek, OR

Joyce Casey, U.S. Forest Service, Washington, DC

Hank Cauley, Forest Stewardship Council, Washington, DC

Susan Chapp, Forestry Action Committee, Cave Junction, OR

Doug Crandall, House Resources Subcommittee on Forests and Forest Health, Washington, DC

Mary Cuolombe, American Forest and Paper Association, Washington, DC

[1] Institutional affiliations date from the time of the interview or workshop. Some have subsequently changed.

Cecilia Danks, Watershed Research and Training Center, Hayfork, CA
Lisa Diehl, National Network of Forest Practitioners, Auburn, WV
Colin Donahue, Rural Action, Trimble, OH
Crockett Dumas, U.S. Forest Service, Ferron, UT
Maia Enzer, Sustainable Northwest, Portland, OR
Gerald Filbin, Environmental Protection Agency, Washington, DC
Kira Finkler, Senate Committee on Energy and Natural Resources, Washington, DC
Douglas Fir, Institute of Sustainable Forestry, Redway, CA
Wayne Fitzpatrick, Forestry Action Committee, Cave Junction, OR
Mike Francis, The Wilderness Society, Washington, DC
Michael T. Goergen, Society of American Forests, Washington, DC
Ken Herrick, Forestry Action Committee, Cave Junction, OR
Steve Holmer, American Lands Alliance, Washington, DC
Sherry Hopper, Forestry Action Committee, Cave Junction, OR
Bill Imbergamo, National Association of State Foresters, Washington, DC
Phil Janik, U.S. Forest Service, Washington, DC
Nels Johnson, World Resources Institute, Washington, DC
Carol Judy, Woodland Community Land Trust, Clairfield, TN
Lynn Jungwirth, Watershed Research and Training Center, Hayfork, CA
Douglas Kenney, Natural Resources Law Center, University of Colorado, Boulder, CO
Faye Knox, Newton County Resource Council, Jasper, AK
Richard Lewis, American Pulpwood Association, Washington, DC
Sungnome Madrone, Redwood Community Action Agency, Eureka, CA
Edwin Marquez, Las Humanas Cooperative, Tajique, NM
Steve Marshall, U.S. Forest Service, Washington, DC
Gary McVicker, Bureau of Land Management, Lakewood, CO
Ruth McWilliams, U.S. Forest Service, Washington, DC
Susan O'Dell, U.S. Forest Service, Washington, DC
Renee Price, Federation of Southern Cooperatives, c/o Land Loss Prevention Project, Durham, NC
George Ramirez, Las Humanas Cooperative, Tajique, NM
Mark Rey, Senate Committee on Energy and Natural Resources, Washington, DC

Jason Rutledge, Healing Harvest Forest Foundation, Copper Hill, VA
Jack Shipley, Applegate Partnership, Grants Pass, OR
Randy Stemler, Mattole River Restoration Council, Petrolia, CA
Rodney Stone, U.S. Forest Service, Baton Rouge, LA
Kirt Taylor, Forestry Action Committee, Cave Junction, OR
Mary Jo Taylor, Forestry Action Committee, Cave Junction, OR
Gus Townes, Alabama Forestry Commission, Montgomery, AL
Don Voth, University of Arkansas, Fayetteville, AR
Jude Waite, Institute of Sustainable Forestry, Redway, CA
Ronnie Yimsut, U.S. Forest Service, Bend, OR

Workshop Participants

Northeast

Russell Barnes, Forestry Contractor, Lyme, NH
Charlie Baylies, Ecosystem-Based Management Consulting Forester, Whitefield, NH
Sam Brown, Forestry Contractor, Parkman, ME
Brian Donahue, Brandeis University, Waltam, MA
Larry Fischer, Cornell University, Ithaca, NY
Jim Heyes, New England Forestry Foundation, Orange, MA
David Kittredge, University of Massachusetts, Amherst, MA
Valerie Luzadis, State University of New York, Syracuse, NY
Brooke Marteins, YellowWood Associates, Inc., St. Albans, VT
Roger Plourde Jr., Consulting Forester, Worster, MA
Hugh Putnam, Consulting Forester, Milton, MA
Michael Snyder, County Forester, Essex Junction, VT
Claudia Swain, New York Watershed Project, Oneonta, NY

Intermountain West

James Burchfield, Bolle Center for People and Forests, University of Montana, Missoula, MT
Sam Burns, Fort Lewis College, Durango, CO
Max Cordova, La Montana de Truchas, Truchas, NM
Carol Daly, Flathead Economic Policy Center, Columbia Falls, MT

John Degroot, Nez Perce Tribal Forestry, Lapwai, ID

Carla Harper, Montezuma County Federal Lands Program, Cortez, CO

Donna House, Ethnobotanist, Indigenous Environmental Issues Consultant, San Juan Pueblo, NM

Ann Moote, University of Arizona, Tucson, AZ

Michael Quintana, Community Forestry Organizer, Chimayo, NM

Lillian Trujillo, La Montana de Truchas, Truchas, NM

Betty Vega, Cooperative Ownership Development Corporation, Silver City, NM

Pacific West

Jenny Blumenstein, Mason County Literacy Program, Shelton, WA

Beverly Brown, Jefferson Center for Education and Research, Wolf Creek, OR

Susan Chapp, Forestry Action Committee, Cave Junction, OR

Sherlette Colegrove, Alliance of Forest Workers and Harvesters, Hoopa, CA

Yvonne Everett, Humboldt State University, Arcata, CA

Wendy George (Poppy), California Indian Basketweaving Association, Basket Weaver, Hoopa, CA

Saoul Guijosa, Nontimber Forest Products Collector, Shelton, WA

Jennifer Kalt, California Indian Basketweaving Association, Willow Creek, CA

Juan Mendoza, Willamette Valley Reforestation, Inc., Molalla, OR

Jose Montenegro, El Centro Internacional Para El Desarrollo Sustentable, Salinas, CA

Bill Otani, U.S. Forest Service, Sandy, OR

Denise Smith, Nontimber Forest Products Collector, Willow Creek, CA

Jude Wait, Institute of Sustainable Forestry, Redway, CA

Notes

Chapter 1

1. Similar and related movements include the environmental justice movement, brownfield redevelopment, the "smart growth" movement, participatory and deliberative city planning (Forester 1999), and the multiplicity of community-based neighborhood and urban renewal groups and organizations.

Chapter 2

1. For more on the debate regarding the extent to which the loss of communal areas contributed to the endemic poverty of the region, allegations that Forest Service grazing permit procedures favored Anglos over Hispanics, and Hispanic resistance, see Carlson 1990 and DeBuys 1985.
2. Although most of these municipal community forests are located in the northeastern United States, others are scattered across the country. A West Coast example is the City of Arcata's community forest. Comprising approximately 600 acres of valuable second-growth redwood- and Douglas fir–dominated stands in two different parcels, Arcata's community forest is managed by the city forester for both intensive recreational use and revenue generation. The sustainability of the forest management regime was attested by a successful certification assessment conducted by the nonprofit SmartWood (an affiliate of the Rainforest Alliance) using the Forest Stewardship Council's guidelines for forest certification.

Chapter 3

1. Using a model based on Ricardian economic theory, Romm argues that the reservation of vast areas of public domain land concentrated people in the remaining unreserved areas. The concomitant closure of frontier areas to settlement and reduction in resource flows and labor opportunities suppressed

wages and increased the wealth of those who own private lands (Romm 2002:122).

2. Romm argues that protecting natural resources by establishing territorially based restrictions on access and race-based social policies that excluded people of color from social opportunities evolved simultaneously and are interlinked. Current attempts to democratize forest management therefore must confront and overcome the legacy of both restricted access and restricted opportunity.
3. For example, "most Black farmers were never told that they could cut 50 acres of marginal land in pine trees and get their loans deferred. In the meantime white farmers were getting Farm Home Administration loans to buy their farms in foreclosure" (Muhammad 1999).
4. Reasons why basic labor issues such as wages and benefits, working conditions, and issues of job safety were not addressed through New Deal era and subsequent programs and policies aimed at achieving community stability in rural resource-dependent communities include the dominant twentieth-century political alignment between government and corporate industrial interests, which resulted in the marginalization of labor issues and a reluctance to champion interests at odds with those of powerful industry interests; the strength of capitalist ideology, which made it easier for antiunion forces to accuse unions of socialist leanings and successfully use red-baiting tactics; and the development of a notion of community within community development discourse (within both academia and policy arenas) that tended to weaken class sensitivities and undermine the importance of working-class issues as opposed to other "community" issues such as infrastructure needs and resource needs of small businesses.
5. Throughout the early twentieth century the long hours, harsh working conditions, lack of benefits, and high accident rates (especially during "speedups") prevalent in logging camps, in mills, and on the docks provided fertile ground for union organizing. Through a variety of intimidation tactics, including violence and running organizers out of town, employer organizations were able to thwart union organizing in the early twentieth century. However, by the 1930s union locals had been successfully established throughout the Pacific Northwest. These included the International Longshoreman's Association, the American Federation of Labor (AFL) Lumber and Sawmill Workers Union, the more militant Congress of Industrial Organizations (CIO) International Longshoremen and Warehousemen's Union, and the CIO International Woodworkers of America. Some of the most important and successful strikes took place in the mid-1930s. During World War II, employer organizations mounted efforts to discredit the union movement as unpatriotic, and after the war, lumber industry employers made concerted efforts to break the power of the unions. Despite this hostile environment, in subsequent decades union locals aggressively sought better wages and benefits and improved safety conditions (Robbins 1988:144–151).

6. This was a period of severe economic depression for the timber industry. In response to low market prices resulting from a slump in market demand, timber industries, locked into competitive market relations, increased timber production. This further compounded the problem of market gluts and low timber prices.
7. When asked by industry to expand its timber sale program in the region of Bend, Oregon, to make up for declining supplies of privately owned old-growth ponderosa pine forests, Silcox refused, stating that he would not allow federal timber to "support such a program of ultimate community disintegration" (quoted in Robbins 1989:15). Instead he supported cooperative forest management ventures with firms that demonstrated their commitment to sustained yield and community stability in part by reducing their mill capacity to sustainable levels.
8. Individual examples belie this general trend and should not be forgotten. For example, the Pennsylvania- and West Coast–based Collins Pine for more than 60 years has practiced socially and ecologically sustainable forest management, although increasingly competitive market pressures challenge their ability to do so. And Pacific Lumber Company (before the junk bond–leveraged hostile corporate takeover by Maxaam Corporation) selectively harvested redwoods in northern California in a manner that observers agreed promoted long-term forest and social sustainability.
9. These factors included the passage of key pieces of environmental legislation such as the Endangered Species Act, the National Forest Management Act, the National Environmental Policy Act, and the Federal Land Policy and Management Act.
10. The forest management gridlock that resulted from the Dwyer and other court decisions led the Clinton administration to convene the 1993 forest conference in Portland, Oregon, to define a scientifically credible compromise solution. The forest ecosystem management assessment team (FEMAT) was responsible for developing an array of options for ecosystem-based forest management. Option 9 was eventually chosen. It established a cut level of 1.2 billion board feet (bbf) per year (significantly lower than the 1980–1989 average cut of 4.5 bbf). (It may have been a hallmark as an idea but not as a reality.)
11. Public participation generally is of limited scope and is controlled closely by the public agency itself. It consists of the following elements. At the beginning of the planning process public input through mailings or hearings is solicited (a process known as scoping) to help identify key issues. In the next phase, plan development (which consists of creating alternative planning scenarios and scenario evaluation criteria, evaluating environmental impacts, and identifying a preferred alternative), there is little or no opportunity for public involvement. The next opportunity for public participation occurs during the formal comment period, when members of the public are able to comment on the draft plan. Public hearings often are also held during the formal comment period. After the close of the 90-day comment pe-

riod, the agency evaluates the comments received, chooses an alternative, and publishes the final plan. At that point the only avenue for further public involvement in the planning process is through administrative appeals and judicial challenges.

Chapter 4

1. One of the authors served on the Board of the Seventh American Forest Congress.
2. The southern United States differs from the Midwest in that it is becoming (again) a primary woodbasket for the country. Corporate forestland owners, facing the declining and increasingly restricted wood supplies of the Pacific Northwest, are attracted to the maturing pine forests of the Deep South and portions of Appalachia. This interest has fueled a consolidation of corporate forestland ownership in states such as Alabama and Arkansas. The increasing nonlocal and large-scale pattern of corporate forestland ownership in these areas has led to concerns that the traditional hunting and fishing rights of local communities and the forest management practices that conserved populations of game and fish will be discontinued and that forest workers will be employed under marginal working conditions with lower pay, all of which are antithetical to the principles of community forestry.
3. Some examples of Lead Partnership Groups are the Applegate Partnership, Jefferson Center, Collaborative Learning Circle, and Forestry Action Committee of Oregon, and the Quincy Library Group, Watershed Research and Training Center, and Shasta–Tehama Bioregional Group of California.
4. This is not to suggest that all Lead Partnership Group pilots have been uniformly successful. The Quincy Library Group–led pilot project, designed primarily to bring national environmental groups into their work, did nothing to reduce the now incendiary relations between the two. However, all-party work continued, with Quincy Library Group members focusing their efforts on monitoring Forest Service work in the Quincy Library Group management plan area.
5. See Colfer and Byron (2001) and Wollenberg et al. (2001). For a comparison of U.S. and international community forestry themes and issues, see Everett and Danks (1996/97).

Chapter 5

1. Of course, the environmental thinking was present long before this time. Henry David Thoreau, George Perkins Marsh, John Muir, Aldo Leopold, and others all contributed vital ideas that informed the late-twentieth-century environmental movement.
2. See Jones et al. (2002) for a recent and comprehensive analysis of these and other nontimber forest product issues.

3. Gaventa et al. (1990), based on studies in the South, suggest that a "technology strategy calls into question the fundamental premise of the trickle-down approach to regional development—industrial development by itself may no longer translate into the creation of jobs and the development of communities." This same technology strategy has been at play in the wood products industry across the country for the last two decades.

Chapter 6

1. A few practitioners of community-based forestry go a step further. In addition to advancing participatory processes, these people and the organizations they represent (e.g., the Jefferson Center and Federation of Southern Cooperatives) explicitly link community forestry with participatory democracy and social justice. They argue that community forestry offers the possibility of reinvigorating American traditions of grassroots-based participatory democracy, within forestry and in broader arenas. Critical analyses of societal oppression combined with a commitment to social justice inform much of their work, which focuses on groups that, for the most part, have not been able to participate in or benefit from the basic political and economic institutions of the United States. The work of these and similar organizations includes popular education, community capacity building, networking, extension and outreach, policy development, improving access to markets, and working to make public institutions (e.g., the U.S. Forest Service and state extension services) responsive to more diverse groups of people. There are striking similarities between the forms of community forestry these groups are advancing and other community movements described as the new social movements. Broadly speaking, the new social movements address issues of political participation, decision-making power, and the democratization of institutions and practices (Young 1990:81). These new social movements demand "that bureaucratic services make possible, instead of replacing, local decision making" (Walzer 1982:152). Young (1990:82) suggests that new social movements "are on the fringes of bureaucratic institutions, . . . carving out new social spaces not dreamt of in their rules, . . . often local and spontaneous, though not unorganized"; they "seek to create alternative institutional forms and independent discussion, . . . exploit and expand the sphere of civil society, . . . and are local and heterogeneous, with loosely networked groups, sharing newsletters or meeting at conferences." Young's characterization of the new social movements rings remarkably true for many of the people and groups in community forestry.
2. The "liberalism" in "interest group liberalism" refers not to liberal ideology but to liberal political theory, which emphasizes the autonomous individuality of people and assumes that individuals are motivated by a set of fixed values and that they act in a rational manner to achieve goals derived from their values.
3. See Foster (2002:140–144) for a similar critique of pluralism within the context of natural resource management.

4. Dryzek buttresses his arguments by drawing on Habermas's distinction between instrumental rationality ("the capacity to devise, select, and effect good means to clarified and consistent ends") and communicative rationality (involving "understanding across subjects, the coordination of their actions through discussion, and socialization"). He suggests (1998:589) that instrumental rationality within liberal democratic political structures, with its emphasis on strategic, calculating, goal-oriented thinking, especially within bureaucracies and capitalist structures, has flourished while communicative rationality has languished. He argues that communicative rationality can provide a model for a more deliberative rather than a strategic democracy.
5. This approach to community forestry has significant implications for the structure, functioning, and role of large government agencies charged with managing public lands. These implications are addressed in Chapter 7.
6. Romm (2000) makes the analogous argument with respect to Progressive Era forest management regimes that, by applying uniform forest policy across diverse conditions, forcefully created homogenous forest conditions that suppressed the preexisting ecological diversity of a region. This process, which dominated the period of forest acquisition, reservation, and management, went hand in hand with the hardening and legal codification of exclusionary modes of racial and class discrimination. From this perspective, community forestry represents a challenge both to the homogenizing legacy of Progressive Era forestry and to the institutionalization of discrimination based on social difference.
7. One example of this from forestry concerns forest workers who are exposed to pesticides and herbicides. In this case the benefits are immediate, but the costs are long term and distributed among an almost invisible group of forest workers. Almost invariably, pesticide applicators are nonlocal, often minority, work crews hired by contractors, often on a seasonal basis; they travel long distances from contract to contract. The mobility of these workers, the seasonality of the work, the relative invisibility of this group of forest workers, and the transitory nature of the workforce make it extremely difficult to track illness related to pesticide exposure. Their invisibility is exacerbated by the fact that they often move through multiple employment sectors in search of seasonal employment (e.g., forestry, agriculture, horticulture, and fisheries). Others who may be exposed to pesticides are planting crews (generally Latino), who move into freshly sprayed areas, and firefighting crews (often Native American) who are unknowingly helicoptered in to recently sprayed remote places. These people, though bearing the majority of the long-term health costs associated with pesticide and herbicide use, are among those least empowered to participate in decision-making processes regarding the distribution of the costs and benefits of chemical use.
8. See the three class action lawsuits filed on behalf of immigrant forest workers against Georgia-Pacific, International Paper, and Champion in Arkansas.

The suits seek redress for widespread violations of wage, overtime, and recordkeeping laws for immigrant workers involved primarily in tree planting (Greenhouse 2001).

9. A Forest Service employee who participated in this project recounted that two perspectives dominate community forestry in the national forests in eastern Oregon: what he calls the "government perspective" and the "private local perspective." Not only do the views, interests, and concerns of nonlocal groups tend to be left out of both perspectives, but the "private local perspective" can be extreme; it has been linked with the county supremacy movement, the occupation of Forest Service land and destruction of public property, and the championing of "local" use over the rights of nonlocal (generally Southeast Asian) forest users, even when "local" users live more than 100 miles away from the forest. This conflict has led to debates within the Forest Service about what is community forestry, who benefits from it, and what is the community within community forestry. Loggers and log truck drivers also may not always be part of the "local" community. Both loggers and truck drivers can travel long distances as part of their seasonal work cycle and therefore may be excluded from place-based community forestry processes. "Place" for loggers is not necessarily the local community or watershed, but instead is the "woodshed" or "workshed."
10. One example of the kinds of challenging conflicts that can arise in these settings is the conflict between subsistence Native American plant harvesting and management and commercial nontimber forest product gathering by Southeast Asian groups. The conflict that arises when these uses are incompatible or negatively affect each other is particularly complex because of the multiple languages, cultural traditions, and systems of customary, treaty, and other legal rights and responsibilities that come into play, especially when both activities occur on public lands.
11. Indeed, in some cases loggers, contractors, and their employees are not only not included in community forestry but are perceived as part of "the problem."
12. An example that drives home the salience of this point, recounted at the Pacific West workshop, concerns an incident in which nontimber forest product gatherers who had gone to the appropriate Forest Service office to obtain harvesting permits were confronted by Immigration and Naturalization Service officers waiting by the back door of the office who asked to see their documentation.
13. An example of this sort of support is the way the Mason County Literacy Program in Shelton, Washington, provides a meeting place for local brush and other nontimber forest product harvesters to gather regularly.
14. The Spanish conference call system is part of what should be a multipronged effort to facilitate the flow of information within and between different communities. The importance of this was stressed at the Pacific West workshop, where it was also argued that much more translation of material from English into other languages is needed.

15. The myth of homogeneity, or of community based on common identity, is based on self-identification as a member of a community rooted in oppositional differentiation from others who may be devalued because they are different. Local autonomy is associated with the ability to exclude others who are different and the ability to exclude or devalue the interests of nonlocal stakeholder groups. Young (1990:250) suggests that autonomy is akin to sovereignty. In this sense autonomy implies closure, the ability to exclude other interests from interfering in what is presumed to be sovereign decision-making authority. It is precisely in local autonomous communities that the dangers of the ideal of community are greatest; "greater local autonomy would be likely to produce even more exaggerated forms of inequalities than current decentralization does." The case of Native American sovereignty is an important exception to the critique of autonomy provided here. Native Americans are autonomous in the sense of having sovereign decision-making authority. This has extremely important implications for the nature of Native American participation in collaborative community forestry processes. They are not another stakeholder but are autonomous governments with their own internal decision-making processes and structures.
16. This effort exemplified the importance and positive results of participatory government processes that recognize the validity and diversity of the needs and values of diverse local groups. The need for federal land management agencies to recognize diverse local perspectives and values and to promote forms of economic development that respect cultural diversity was discussed at the Pacific Northwest workshop. The implications of embracing the politics of difference for government agencies and organizations and for the practice of science are profound.

Chapter 7

1. In this section we focus on the Forest Service because of its central role in community forestry. The Forest Service is a key community forestry player because of the extensive public lands it controls, the large number of communities and community groups with relationship to that land, and its commitment to and support of urban community forestry programs.
2. Perhaps indicative of the tenuous nature of Forest Service involvement with rural communities and workers, President Bush's proposed Fiscal Year 2003 budget seeks to completely eliminate the suite of programs included with the Economic Action Program (Rural Community Assistance, Forest Products Conservation and Recycling, and Market Development and Expansion). From the perspective of community forestry practitioners and supporters, zeroing out the Economic Action Program would eliminate some of the most effective tools communities and the Forest Service have for working together on efforts to build community capacity, support economic diversification, and foster the development of a forest stewardship–based

economy. Contrary to claims by the administration that the U.S. Department of Agriculture Rural Development program will step in to replace the Economic Action Program, supporters of community forestry have noted that the Rural Development program tends to focus on infrastructure, not on building community capacity.

3. See Qi et al. (1998) for a concise review of the history and accomplishments of the Urban and Community Forestry Program.
4. It is important to stress the diversity among the 50 different state forestry organizations in terms of both mandated authorities and tasks and the desire and ability to embrace new program ideas and goals (e.g., community-based ecosystem management). Some state forestry organizations in the southern United States have been reluctant to partner with the Rural Community Assistance program because of the challenging issues associated with low literacy rates and community capacity and related financial management issues. Some southern state foresters have criticized Forest Service programs for being too small and for their minimal landscape-level impact. Meanwhile, other state foresters (e.g., those in Minnesota and New Mexico) have been on the cutting edge of community collaboration and partnering.
5. In his classic study of the Tennessee Valley Authority (TVA), Selznick (1949/1984) analyzed the problems of co-optation that arise when a federal program whose objective is to serve underrepresented groups is implemented through state agencies and land grant universities that are almost invariably closely linked with regionally dominant interests and groups. In the case of the TVA, "grassroots democracy," defined as working with counterparts at the state and regional levels, resulted in a dramatic redirection of federal funds and support away from the intended beneficiaries—small and minority landowners—toward large landowners and capital-intensive agriculturalists.
6. See Fentz et al. (1999) for a fuller analysis of these ideas.
7. See Kusel and Adler (2001), Richard and Burns (1998), and Bernard and Young (1997).
8. In this vein Dumas stated that "the biggest barrier to community forestry is the people who are in the line officer positions, bureaucrats with the authority to make things happen, but don't; folks who are stuck in the old 1978 forest planning mode in which neither employees nor the public were truly engaged in the planning process." Others have also critiqued the "1978 planning mode" for a variety of reasons (see related discussion in Chapter 3).
9. The organizational models that Dumas, McVicker, and others are developing from the ground up correlate closely with more theoretically derived formulations regarding the organizational and institutional implications of ecosystem management. For example, Meidinger (1997:370) shows why "loosely coupled networks" as opposed to hierarchical organizations that emphasize top-down information flows are appropriate for ecosystem man-

agement. An organization that functions more like a network than a hierarchy devotes resources to developing a shared understanding of organizational mission and identity (both internally and in conjunction with its civic partners), achieves coordination "through regular and informal mutual adjustment," and substitutes information exchange for supervision and control. Meidinger goes on to contrast the role of bureaucracy under pluralism (the distribution of benefits to the strongest interest group) with civic republicanism (the development of a shared understanding of the common good through public deliberation). Under civic republicanism, organizations promote deliberation, "resist hierarchy," and have a slight bias toward the local. This requires bureaucratic models that allow "extensive communication of knowledge and values among citizens as they solve the problems of their collective life." This model of organizational behavior closely parallels the organizational forms that Dumas, McVicker, and others have evolved by innovating, taking risks, and experimenting with new approaches to old problems.

10. The barriers and challenges associated with institutionalizing the changes in the Forest Service and BLM necessary for the long-term support of community forestry are common to most forest departments around the world that are struggling to embrace community forestry. Edmunds and Wollenberg (2001:192–197) review many of these barriers and challenges in the context of forest management in Asia; many of the issues they discuss are directly analogous to issues in the United States. They include the widespread belief among professional foresters that local communities possess neither the technical nor the managerial skills necessary for forest management, arguments from the conservation (i.e., environmental) community that local control will threaten the public interest in forests, collaborative and participatory programs that devolve very little decision-making authority and in some cases shift forest management burdens onto local communities without providing commensurate benefits or substantial power sharing, the ability of forest department staff to impede progress of community forestry programs through their control of the bureaucratic process and information flows, links between local elites or other dominant groups and the forest department that result in community forestry programs and policies that work against the interests of rural marginalized communities and groups, and situations in which local institutional, financial, technical, and social capacities for forest management have been reduced or weakened by historical processes that have prevented their exercise.
11. See Frentz et al. (1999) for a comprehensive set of recommendations regarding what sorts of changes are needed for the Forest Service to more fully embrace community forestry.
12. See Ringgold (1998) for a good description of the diverse array of contracting authorities and procedures currently available to the Forest Service. Although a full discussion and assessment of these diverse programs are be-

yond the scope of this book, it is worth noting that the extent of innovation is significant and, if monitoring and evaluation are successful, should provide useful insights into the sorts of programs that are likely to be most successful. Great effort is currently directed toward monitoring these innovations. For example, the Pinchot Institute has been contracted by the Forest Service to monitor its 28 pilot stewardship projects, and member organizations of the Communities Committee have been working with the Forest Service and other government representatives to help ensure thorough monitoring of the social, ecological, and economic effects of the implementation of the fire plan.

13. In fact, in some cases current trends suggest a movement away from community forestry, at least in some parts of the Forest Service and in some geographic locations. For example, in some Arkansas rural communities adjacent to national forests, local contracts that had been previously let to local community businesses and individuals are now awarded to nonlocal contractors. This has resulted in bitterness in the local business community against the Forest Service. A similar point is that in some cases, because of persistent budget cuts on one hand and calls to address the fire and forest restoration issues in the West on the other, Forest Service officials are scrambling to get as much work done as cheaply as possible; there is little time, money, and organizational space to experiment with new approaches, even if over the long run they may turn out to be more cost effective.
14. Even in regions where extension forestry may not embrace community forestry, there are often innovative extension foresters who focus their efforts on promoting community forestry and working with nontraditional client groups. The challenge, as always, is to create a supportive institutional environment that is conducive to community forestry.
15. Many researchers and supporters of community forestry around the world have argued for the importance of local mobilization to demand effective implementation of formal devolution and collaboration-oriented policies. Edmunds and Wollenberg (2001:197) note that "while certainly not a sufficient condition, local mobilization may be necessary to make devolution reach its most democratic forms." In a similar vein, Gilmour and Fisher (1997:17, cited in Edmunds and Wollenberg 2001:197) argue that "unless community based natural resource management initiatives develop into broadly based social movements, they are unlikely to be politically and institutionally sustainable in the long term."
16. In an interesting divergence from the trend at the national level, environmentalists who live near forestlands often support limited forest restoration activity focused on fuel reduction and forest thinning. This support for local forms of collaborative resource management by local environmentalists has kicked off debates and resulted in tensions between them and national environmental groups that oppose such forms of resource management.
17. Two representatives of national environmental groups based in Washington,

D.C., when asked to define community-based forestry, had the following to say about it: "Local vested economic interests [are] gaining control of national assets through community-based strategies," and "community-based forestry is a subsidized attempt by industry to turn things in their favor and as a way to get around what professional managers say should be done with our national lands. The Quincy Library Group is an example of the worst of this, and the Applegate Partnership a better example." It's unclear what this representative meant by "worst" and "better."

18. John (1994) echoes this theme in his book *Civic Environmentalism,* in which he raises serious doubts about the ability of traditional command-and-control regulation to address the "unfinished business" of environmental management, which, he argues, consists of nonpoint source pollution, pollution prevention, and ecosystem management. Based on three in-depth case studies and the analysis of secondary data, John shows that the "unfinished business" of environmental management is characterized by unprecedented knowledge requirements, local variations, ecological complexity, a large number of players, and risk and uncertainty. These characteristics make command-and-control regulation blunt and ineffective, so they necessitate a different approach, one that John calls civic environmentalism.
19. There are good reasons why community forestry is likely to be significantly more ecologically beneficial than the range of activities that comply with legislative and other guidelines and restrictions. The first is community forestry's primary focus on reinvestment and ecological restoration. A key thrust of community forestry is to reverse the historical patterns of long-term disinvestment that have pauperized both natural and social systems. The widespread embrace of Aldo Leopold's land ethic and philosophy of land stewardship by the community forestry movement symbolizes this intent. Innovative mechanisms for promoting various forms of capital reinvestment in natural and social systems, delineated in Chapter 9, are actualizing the intent. A second reason is that most of the resource extraction practices community forestry promotes are small-scale and localized, use site-specific knowledge, emphasize efficiency of resource product use, are experimental, and conform to practices associated with adaptive management. Extractive activities with these characteristics are inherently conservative. Based on cutting-edge scientific and folk knowledge, they are designed to help increase ecological diversity and resiliency, promote ecological productivity, and repair damage associated with historic patterns of resource degradation. A third reason stems from the relationships between local communities and the surrounding natural environment. Much of the historical deforestation in the United States is associated with the activities of regional or national timber companies and corporations for whom short-term profits, executive salaries, quarterly reports, shareholder concerns, and the prospect that cutover lands could always be sold and uncut areas purchased were more important than long-term ecological

stewardship. In contrast, forest product extraction and use practices associated with community forestry are designed and implemented primarily by local people and businesses for whom the vision of long-term ecological health and environmental stewardship is a driving principle. This is underscored by the fact that one of the most widely held values of community forestry practitioners is to be able to bequeath to their children and grandchildren a natural environment of equal if not better health and productivity than currently exists. This sentiment stems from values associated with place-based attachments.

20. The Wilderness Society and the Nature Conservancy are among the national environmental organizations that are seeking to constructively engage with the community forestry movement through the ideas, methods, and approaches of "community-based conservation." Other environmental groups are going through periods of lively internal debate regarding whether and how to constructively work with community forestry.
21. Forest arson rates have increased in areas where customary hunting and fishing rights have been extinguished.
22. The nonrenewal of the U.S.–Canadian softwood agreement exemplifies this point.

Chapter 8

1. Epistemology is defined as the study or theory of the nature and grounds of knowledge, especially with reference to its limits and validity. Therefore, at the heart of the methods, culture, and worldview of science lies an epistemology that is often unexamined by scientists and nonscientists alike. For example, scientific objectivity is predicated on the epistemological position that the "reality" lying behind our observations can be progressively elucidated by inquiry, investigation, and experimentation.
2. Science historians and philosophers call traditional science *normal* science.
3. As mentioned previously, lessons brought home from work in developing countries, particularly those in Africa and Latin America, have had a profound effect. Robert Chambers's (1997) work, with a particular concern for equity, among other issues, is one of the more notable in its effect on U.S. resource social scientists.

Chapter 9

1. Sixty percent of that total is the value of water. Other commodities account for 20 percent, and services account for another 20 percent (Stewart 1996:974).
2. This categorization and discussion are drawn primarily from Luzadis et al. (2001). It is supplemented by ideas contained in Best (2000).
3. To date, certified wood, for the most part, has not generated a premium

price. Nonetheless, forest certification can bring with it a whole suite of other benefits. Some of the benefits associated with Forest Stewardship Council certification, for example, include support for responsible management of nontimber forest products; opportunities for employment, training, and other services for local communities and investment in the local economy; support for efforts to retain the economic benefits of forest management within local economies; standards that promote high-quality work opportunities for all workers; and protection for the rights of workers to organize and negotiate with their employers. Many of these benefits are entirely consistent with community forestry objectives.

4. Luzadis et al. (2001:176) describe The Nature Conservancy's Clinch Valley Forest Bank, created in 1998, as one example of a resource bank. Citing Mater (1997), they suggest that the interest landowners receive on their principal would equal or exceed what landowners would receive by harvesting the trees themselves.
5. Investments that focus exclusively on income improvement strategies for individuals have numerous drawbacks. They may exacerbate "brain drain" and capital flight from the poorest places. Also, when poverty policies provide incentives for human and economic relocation from chronically poor places to "growth" areas with more economic opportunities, they may have unintended negative consequences for the communities and landscapes left behind.
6. California State Treasurer Philip Angelides observed that "Sustained economic success in the 21st century will require investment of public resources. . . . California will not achieve economic success in the long run if our environment is degraded or if there are pockets of economic failure throughout the State. . . . Taxpayers' dollars should be focused on rebuilding older, decaying cities and other at-risk 'declining communities' instead of creating more far-flung suburbs" (News Release, Press Advisory, June 23, 1999).
7. Carl Anthony (Ford Foundation program officer, former director of urban habitat) links disinvestment, environmental degradation, and environmental racism. He describes urban multicultural populations as "economically and culturally marginalized . . . environmental refugees [who are] uprooted from the land" (Anthony 1998:4).
8. The study goes on to suggest that, "based on 1994 water prices, less than a 10 percent increase in water rates for both agricultural water users ($1.50 per acre-foot price increase) and urban users ($35 per acre-foot price increase) would produce the same amount of revenue without placing large burdens on any type of user" (Standiford et al. 1994:44).
9. The term *impaired* means that a water body does not meet state or federal water quality standards under the Clean Water Act. The standards include turbidity, levels of chemicals, nutrients, sediment, and temperature.
10. In the spring 2001 issue of *Communities and Forests,* the newsletter of the Communities Committee of the Seventh American Forest Congress, Michi-

gan State University foresters working with low-income multicultural communities in Detroit report, "The MSU team expected participants to say that funding was the most important factor for success, they didn't. Instead they identified technical assistance and the empowerment of local residents as the key to their success. . . . These new skills are now spreading to other neighborhoods and cities" (Shepherd 2001:6).

References

Ahl, Valerie, and T. F. H. Allen. 1996. *Hierarchy Theory: A Vision, Vocabulary and Epistemology.* New York: Columbia University Press.

Allen, Jeff. 1992. *Liquid Assets: The Potential of Water User Fees.* Unpublished paper for the California Policy Center and the California Senate Office of Research.

Anderson, Judith L. 1998. "Embracing Uncertainty: The Interface of Bayesian Statistics and Cognitive Psychology." *Conservation Ecology* 2(1): 2. URL: http://www.consecol.org/vol2/iss1/art2.

Anderson, Kat. 1993. "Native Californians as Ancient and Contemporary Cultivators," in *Before the Wilderness: Environmental Management by Native Californians,* Thomas Blackburn and Kat Anderson (eds.), pp. 151–174. Menlo Park, California: Ballena Press.

Angelides, Philip. 1999. Press Advisory. June 23. Office of the Treasurer. Sacramento: State of California.

Anthony, Carl. 1998. *Urban Habitat Program Report* 4.

B.C. Forest Service. 1999. *An Introductory Guide to Adaptive Management for Project Leaders and Participants.* Online report from the Forest Practice Branch, B.C. Forest Service. URL: http://www.for.gov.bc.ca/hfp/amhome/INTROGD/Toc.htm.

Beck, Ulrich. 1995. *Ecological Enlightenment.* Atlantic Highlands, New Jersey: Humanities Press.

Bellah, Robert. 1995. "The Quest for Self," in *Rights and the Common Good,* Amitai Etzioni (ed.), pp. 45–57. New York: St. Martin's Press.

Bellah, Robert, Richard Madsen, William Sullivan, Ann Swidler, and Steven Tipton. 1991. *The Good Society.* New York: Alfred A. Knopf.

Berkes, Fikret. 1999. *Sacred Ecology: Traditional Ecological Knowledge and Resource Management.* Washington, D.C.: Taylor & Francis.

Bernard, Ted, and Jora Young. 1998. *The Ecology of Hope: Communities*

Collaborate for Sustainability. Gabriola Island, B.C.: New Society Publishers.

Best, Constance. 2000. "Rebuilding Forest Ecosystem Assets by Better Aligning Values, Markets, and Rights." Paper prepared for the conference "Natural Assets: Democratizing Environmental Ownership," January 21–23, Santa Fe, New Mexico.

Blackburn, Thomas, and Kat Anderson (eds.). 1993. *Before the Wilderness: Environmental Management by Native Californians.* Menlo Park, California: Ballena Press.

Bliss, John, Greg Aplet, Cate Hartzell, Peggy Harwood, Paul Jahnige, David Kittredge, Stephen Lewandowski, and Mary Sue Sascia. 2001. "Community Based Ecosystem Monitoring," in *Understanding Community-Based Forest Ecosystem Management,* Gerald Gray, Maia Enzer, and Jonathan Kusel (eds.), pp. 143–167. New York: Haworth Press.

Blumberg, Louise, and Darrell Knuffke. 1998. "Count Us Out: Why the Wilderness Society Opposed the Quincy Library Group Legislation." *Chronicle of Community* 2(2): 41–44.

Borchers, Jeffrey G. 1996. "A Hierarchical Context for Sustaining Ecosystem Health," in *Search for a Solution: Sustaining the Land, People, and Economy of the Blue Mountains,* Raymond G. Jaindl and Thomas M. Quigley (eds.), pp. 63–80. Washington, D.C.: American Forests Publications.

Boyce, James K., and Berry G. Shelly. Forthcoming. *Natural Assets: Democratizing Environmental Ownership.* New York: Russel Sage.

Bradshaw, Gay A. 1998. "Defining Ecologically Relevant Change in the Process of Scaling-Up: Implications for Monitoring at the Landscape Level," in *Ecological Scales: Theory and Applications,* David L. Peterson and V. Thomas Parker (eds.), pp. 227–249. New York: Columbia University Press.

Bradshaw, Gay A. 2001. "Ecology and Social Responsibility: The Re-Embodiment of Science." *Trends in Ecology and Evolution* 16(8): 460–465.

Bradshaw, Gay A., and Jeff G. Borchers. 2000. "Narrowing the Science–Policy Gap: Uncertainty as Information." *Conservation Ecology* 4(1): 7.

Brendler, Thomas, and Henry Carey. 1998. "Community Forestry, Defined." *Journal of Forestry* 96(3): 21–23.

Brick, Philip. 1998. "Of Imposters, Optimists, and Kings: Finding a Political Niche for Collaborative Conservation." *Chronicle of Community* 2(2): 34–38.

Brown, Beverly. 2001. "Integrating Forest Labor Participation into Community-Based Ecosystem Management Processes," in *Understanding Community-Based Forest Ecosystem Management,* Gerald Gray, Maia Enzer, and Jonathan Kusel (eds.), pp. 291–304. Binghamton, New York: Haworth Press.

Brown, Lee, and Helen Ingram. 1987. *Water and Poverty in the Southwest.* Tucson: University of Arizona Press.

Bunker, Stephen. 1985. *Underdeveloping the Amazon Extraction, Unequal Exchange, and the Failure of the Modern State.* Urbana: University of Illinois Press.

Carlson, Alvar. 1990. *The Spanish–American Homeland: Four Centuries in New Mexico's Rio Arriba.* Baltimore: Johns Hopkins University Press.

Carpenter, Stephen, William Brock, and Paul Hanson. 1999. "Ecological and Social Dynamics in Simple Models of Ecosystem Management." *Conservation Ecology* 3(2): 4. URL: http://www.consecol.org/vol3/iss2/art4.

Caucus of Regulated Representatives. 2001. *AB 982 PAG Regulated Caucus Comments Regarding TMDL Initiative and Action Plan.* Sacramento, California. Unpublished letter.

Chambers, Robert. 1997. *Whose Reality Counts? Putting the Last First.* London: Intermediate Technology Publications.

Cheek, Kristin. 1996. *Community Well-Being and Forest Service Policy: Re-Examining the Sustained Yield Unit.* M.S. thesis, Oregon State University.

Cilliers, Paul. 1998. *Complexity and Post-Modernism: Understanding Complex Systems.* London: Routledge.

Clary, David. 1987. "What Price Sustained Yield? The Forest Service, Community Stability, and Timber Monopoly Under the 1944 Sustained-Yield Act." *Journal of Forest History* 4(31): 4–18.

Coggins, George. 1999. "Regulating Federal Natural Resources: A Summary Case Against Devolved Collaboration." *Ecology Law Quarterly* 25(4): 602–610.

Colfer, Carol, and Yvonne Byron (eds.). 2001. *People Managing Forests: The Links Between Human Well-Being and Sustainability.* Washington, D.C.: Resources for the Future/Center for International Forestry Research (CIFOR).

Condon, Patrick, and Stacy Moriarty. 1999. *Second Nature: Adapting LA's Landscape for Sustainable Living.* Los Angeles: Metropolitan Water District of Southern California.

Cortner, Hanna J., and Margaret A. Moote. 1999. *The Politics of Ecosystem Management.* Washington, D.C.: Island Press.

Cowan, George A., David Pines, and David Meltzer (eds.). 1994. "Complexity: Metaphors, Models, and Reality." Proceedings Volume XIX, Santa Fe Institute, *Studies in the Sciences of Complexity.* Reading, Massachusetts: Addison-Wesley.

Cranor, Carl F. 1997. "The Normative Nature of Risk Assessment: Features and Possibilities." *Risk* 8: 123.

Cronon, William. 1983. *Changes in the Land: Indians, Colonists, and the Ecology of New England.* New York: Hill & Wang.

Cronon, William. 1991. *Nature's Metropolis: Chicago and the Great West.* New York: W. W. Norton.

Dallman, Suzanne, and Tom Piechota. 1999. *Storm Water: Asset Not Liability.* Los Angeles: The Los Angeles and San Gabriel Rivers Watershed Council.

Dana, Samuel, and Sally Fairfax. 1956/1980. *Forest and Range Policy.* New York: McGraw-Hill.

DeBuys, William. 1985. *Enchantment and Exploitation: The Life and Hard Times of a New Mexico Mountain Range.* Albuquerque: University of New Mexico Press.

Dryzek, John. 1998. "Political and Ecological Communication," in *Debating the*

Earth: The Environmental Politics Reader, John Dryzek and David Schlosberg (eds.), pp. 584–597. New York: Oxford University Press.
Edmunds, David, and Eva Wollenberg. 2001. "Historical Perspectives on Forest Policy Change in Asia: An Introduction." *Environmental History* 6(2): 190–212.
Etzioni, Amitai. 1994. *The Spirit of Community: The Reinvention of American Society.* New York: Simon & Schuster.
Everett, Yvonne, and Cecilia Danks. Winter 1996/1997. "Community Based Forest Management: International Lessons Applied in the Trinity Bioregion of Northern California, USA." Rural Development Forestry Network Paper 20a. London: ODI, Portland House.
Fischer, Frank. 1990. *Technocracy and the Politics of Expertise.* Newbury Park, California: Sage.
Fischer, Frank. 2001. *Citizens, Experts, and the Environment: The Politics of Local Knowledge.* Durham, North Carolina: Duke University Press.
Forester, John. 1999. *The Deliberative Practitioner: Encouraging Participatory Planning Processes.* Cambridge, Massachusetts: MIT Press.
Fortmann, Louise, Jonathan Kusel, and Sally Fairfax. 1989. "Community Stability: The Foresters' Figleaf," in *Community Stability in Forest-Based Economies: Proceedings of a Conference in Portland, OR, November 16–18, 1987,* Dennis LeMaster and John Beuter (eds.), pp. 44–50. Portland, Oregon: Timber Press.
Foster, Sheila. 2002. "Environmental Justice in an Era of Devolved Collaboration," in *Justice and Natural Resources: Concepts, Strategies and Applications,* Kathryn Mutz, Gary Bryner, and Douglas Kenney (eds.), pp. 139–160. Washington, D.C.: Island Press.
Francis, R. I. C. C., and R. Shotton. 1997. "'Risk' in Fisheries Management: A Review." *Canadian Journal of Fisheries and Aquatic Science* 54: 1699–1715.
Frentz, Irene, Sam Burns, Donald Voth, and Charles Sperry. 1999. *Rural Development and Community-Based Forest Planning and Management: A New Collaborative Paradigm.* Fayetteville: University of Arkansas Agricultural Experiment Station.
Funtowicz, Silvio, and Jerome R. Ravetz. 1993. "Science for the Post-Normal Age." *Futures* 25: 735–755.
Funtowicz, Silvio, and Jerome R. Ravetz. 1999. "Post-Normal Science: An Insight Now Maturing." *Futures* 31: 641–646.
Funtowicz, Sylvio, J. Martinez-Alier, G. Munda, and Jerome R. Ravetz. 1999. Information Tools for Environmental Policy Under Conditions of Complexity. Environmental issue report No. 9. Luxembourg: Office for Official Publications of the European Communities.
GAO. 1997. *Forest Service Decision Making: A Framework for Improving Performance.* GAO/T-RECD-97-153, April 29. Washington, D.C.
Gaventa, John. 1980. *Power and Powerlessness: Quiescence and Rebellion in an Appalachian Valley.* Urbana: University of Illinois Press.
Gaventa, John. 1993. "The Powerful, the Powerless, and the Experts: Knowl-

edge in the Information Age," in *Voices of Change: Participatory Research in the United States and Canada,* Peter Park, Mary Brydon-Miller, Bud Hall, and Ted Jackson (eds.), 21–40. Toronto: OISE Press.

Gaventa, John, B. E. Smith, and A. Willingham. 1990. "Toward a New Debate: Development, Democracy, and Dignity," in *Communities in Economic Crises: Appalachia and the South,* John Gaventa, Barbara Smith, and Alex Willingham (eds.), 279–291. Philadelphia: Temple University Press.

Gilmour, Donald, and Robert Fisher. 1997. "Evolution in Community Forestry: Contesting Forest Resources." Paper presented at the International Seminar on Community Forestry at the Crossroads: Reflections and Future Directions in the Development of Community Forestry, Bangkok, July 17–19.

Goergen, Michael, Donald Floyd, and Peter Ashton. 1997. "An Old Model for Building Consensus and a New Role for Foresters." *Journal of Forestry* 95(1): 8–12.

Gould, Stephen. J. 2000. "Deconstructing the 'Science Wars' by Reconstructing an Old Mold." *Science,* January 14: 253–261.

Gray, Gerald, Maia Enzer, and Jonathan Kusel. 2001. Synthesis paper in *Understanding Community-Based Forest Ecosystem Management.* New York: Haworth Press.

Greenhouse, Steven. 2001. "Migrants Plant Pine Trees but Often Pocket Peanuts." *New York Times,* February 14, p. A16.

Gunderson, Lance H., and Crawford S. Holling. 2001. *Panarchy: Understanding Transformations in Systems of Humans and Nature.* Washington, D.C.: Island Press.

Gunderson, Lance H., Crawford S. Holling, and Stephen S. Light (eds.). 1995. *Barriers and Bridges to the Renewal of Ecosystems and Institutions.* New York: Columbia University Press.

Hays, Samuel. 1959. *Conservation and the Gospel of Efficiency.* Cambridge, England: Harvard University Press.

Hempel, Lamont. 1999. "Conceptual and Analytical Challenges in Building Sustainable Communities," in *Towards Sustainable Communities: Transition and Transformation in Environmental Policy,* Daniel Mazmanian and Michael Kraft (eds.), pp. 43–74. Cambridge, Massachusetts: MIT Press.

Hess, Karl, Jr. 1996. "Wising Up to the Wise Use Movement," in *A Wolf in the Garden: The Land Rights Movement and the New Environmental Debate,* Philip Brick and Cawley McGregor (eds.), pp. 161–184. Lanham, Maryland: Rowman & Littlefield.

Higgs, Eric S. 1997. "What Is Good Ecological Restoration?" *Conservation Biology* 11: 338–348.

Hiss, Tony. 1990. *The Experience of Place.* New York: Alfred A. Knopf.

Holling, C. S. (ed.). 1978. *Adaptive Environmental Assessment and Management.* New York: Lang and Wiley.

John, Dewitt. 1994. *Civic Environmentalism: Alternatives to Regulation in States and Communities.* Washington, D.C.: Congressional Quarterly Press.

Johnson, Nels, Andy White, and Daniele Perrot-Maitre. 2001. *Developing Markets for Water Services from Forests: Issues and Lessons for Innovators.* Washington, D.C.: Forest Trends.

Jones, Eric, Rebecca McLain, and James Weigand. 2002. *Nontimber Forest Products in the United States.* Lawrence: University of Kansas Press.

Kemmis, Daniel. 1996. *Community and the Politics of Place.* Norman: University of Oklahoma Press.

Kemmis, Daniel. 2001. *This Sovereign Land: A New Vision for Governing the West.* Washington, D.C.: Island Press.

Kenney, Douglas, Sean McAllister, William Caile, and Jason Peckham. 2000. *The New Watershed Source Book.* Boulder, Colorado: Natural Resources Law Center.

Krahl, Lane, and Doug Henderson. 1998. "Uncertain Steps Toward Community Forestry: A Case Study in Northern New Mexico." *Natural Resources Journal* 38: 53–84.

Kuhn, Thomas S. 1962. *The Structure of Scientific Revolutions.* Chicago: University of Chicago Press.

Kusel, Jonathan. 1991. "Ethnographic Analysis of Three Forest Communities in California," in Vol. 2 of *Well-Being in Forest Dependent Communities.* Sacramento: California Department of Forestry, Forest and Rangeland Assessment Program.

Kusel, Jonathan. 1996. "Well-Being in Forest Dependent Communities, Part I: A New Approach," in Sierra Nevada Ecosystem Project: Final Report to Congress, Volume II, *Assessment and Scientific Basis for Management Options,* pp. 361–374. Davis: University of California, Centers for Water and Wildland Resources.

Kusel, Jonathan. 2001. "Assessing Well-Being in Forest Dependent Communities." *Journal of Sustainable Forestry* 13(1/2): 359–384.

Kusel, Jonathan, et al. 2002. *Assessment of the Northwest Economic Adjustment Initiative.* Taylorsville, California: Forest Community Research.

Kusel, Jonathan, and Elisa Adler (eds.). 2003. *Forest Communities, Community Forests: A Collection of Case Studies of Community Forestry.* Lanham, MD: Rowan and Littlefield Publishers, Inc.

Kusel, Jonathan, Sam C. Doak, Susan Carpenter, and Victoria E. Sturtevant. 1996. "The Role of the Public in Adaptive Ecosystem Management," in Sierra Nevada Ecosystem Project: Final Report to Congress, Volume II, *Assessments and Scientific Basis for Management Options,* pp. 611–624. Davis: University of California, Centers for Water and Wildland Resources.

Kusel, Jonathan, Lee Williams, and Diana Keith. 2000. *A Report on All-Party Monitoring and Lessons Learned from the Pilot Projects.* Forest Community Research and the Pacific West National Community Forestry Center, Technical Report 101–2000. Taylorsville, California.

Lavigne, Peter. 2003. "Forestry at the Urban–Rural Interface: The Beaver Brook Association and the Merrimack River Watershed," in *Forest Communities, Community Forests: A Collection of Case Studies in Community Forestry,*

Jonathan Kusel and Elisa Adler (eds.), pp. 199–216. Lanham, MD: Rowan and Littlefield Publishers, Inc.

Lee, Eliza. 1995. "Political Science, Public Administration, and the Rise of the American Administrative State." *Public Administration Review* 55(6): 538–546.

Lee, Kai N. 1993. *Compass and Gyroscope: Integrating Science and Politics for the Environment.* Washington, D.C.: Island Press.

Lee, Kai N. 1999. "Appraising Adaptive Management." *Conservation Ecology* 3(2): 3. URL: http://www.consecol.org/vol3/iss2/art3.

Lee, Robert, Donald Field, and William Burch. 1989. *Community and Forestry: Continuities in the Sociology of Natural Resources.* Boulder, Colorado: Westview Press.

Lewis, Henry. 1993. "Patterns of Indian Burning in California: Ecology and Ethnohistory," in *Before the Wilderness: Environmental Management by Native Californians,* Thomas Blackburn and Kat Anderson (eds.), 55–116. Menlo Park, California: Ballena Press.

Limerick, Patricia. 2002. "Hoping Against History: Environmental Justice in the Twenty-First Century," in *Justice and Natural Resources: Concepts, Strategies and Applications,* Kathryn Mutz, Gary Bryner, and Douglas Kenney (eds.), pp. 337–354. Washington, D.C.: Island Press.

Lindquist, Donna, and Jim Wilcox. 1999. "New Concepts for Meadow Restoration in the Northern Sierra Nevada." Unpublished paper.

Lubchenco, Jane. 1998. "Entering the Century of the Environment: A New Social Contract for Science." *Science* 279: 491–497.

Luzadis, Valerie, Carolyn Alkire, Catherine Mater, Jeff Romm, William Stewart, Leah Wills, and Duane Vaagen. 2001. "Investing in Ecosystems and Communities," in *Understanding Community-Based Forest Ecosystem Management,* Gerald Gray, Maia Enzer, and Jonathan Kusel (eds.), pp. 169–194. Binghamton, New York: Haworth Press.

MacKaye, Benton. 1918. "Some Social Aspects of Forest Management." *Journal of Forestry* 16(2): 210–214.

MacKaye, Benton. 1919. "Employment and Natural Resources: Possibilities of Making New Opportunities for Employment through the Settlement and Development of Agricultural and Forest Lands and Other Resources." Department of Labor. Washington, D.C.: General Printing Office.

Marchak, Patricia. 1990. "Forest Industry Towns in British Columbia," in *Community and Forestry: Continuities in the Sociology of Natural Resources,* Robert Lee, Donald Field, and William Burch (eds.), pp. 95–106. Boulder, Colorado: Westview Press.

Marcot, Bruce G. 1998. "Selecting Appropriate Statistical Procedures and Asking the Right Questions: A Synthesis," in *Statistical Methods for Adaptive Management Studies,* Vera Sit and Brenda Taylor (eds.), pp. 129–141. Victoria: Research Branch, B.C. Ministry of Forests, Land Management Handbook No. 42.

Mason, David. 1927. "Sustained Yield and American Forest Problems." *Journal of Forestry* 25: 625–658.

Mater, Catherine. 1997. *The Nature Conservancy Clinch Valley Forest Bank Business Plan.* Corvallis, Oregon: Mater Engineering, Ltd.

Mazmanian, Daniel, and Michael Kraft. 1999. *Toward Sustainable Communities: Transition and Transformations in Environmental Policy.* Cambridge, Massachusetts: MIT Press.

McCarthy, John, and Mayer Zald. 1977. "Resource Mobilization and Social Movements: A Partial Theory." *American Journal of Sociology* 82(6): 1212–1241.

McCloskey, Michael. 1999. "Local Communities and the Management of Public Forests." *Ecology Law Quarterly* 25(4): 624–629.

McCullough, Robert. 1995. *The Landscape of Community: A History of Communal Forests in New England.* Hanover, New Hampshire: University Press of New England.

McIntosh, Robert P. 1986. *The Background of Ecology: Concept and Theory.* New York: Cambridge University Press.

Meidinger, Errol. 1997. "Organizational and Legal Challenges for Ecosystem Management," in *Creating a Forestry for the 21st Century: The Science of Ecosystem Management,* Kathryn Kohm and Jerry Franklin (eds.), pp. 361–379. Washington, D.C.: Island Press.

Mitsos, Mary. 2003. "Western Upper Peninsula Forest Improvement District, Michigan: Adding Value to a Working Landscape," in *Forest Communities, Community Forests: A Collection of Case Studies in Community Forestry,* Jonathan Kusel and Elisa Adler (eds.), pp. 151–162. Lanham, MD: Rowan and Littlefield Publishers, Inc.

Moote, Margaret A., and Mitchell McClaran. 1997. "Implications of Participatory Democracy for Public Land Planning." *Journal of Range Management* 50(5): 473–481.

Morgan, M. Granger, and Max Henrion. 1990. *Uncertainty: A Guide to Dealing with Uncertainty in Quantitative Risk and Policy Analysis.* New York: Cambridge University Press.

Moseley, Cassandra, and Staccy Shamkle. 2001. "Who Gets the Work? National Forest Contracting in the Pacific Northwest." *Journal of Forestry* 99(9): 32–37.

Muhammad, Ridgely A. Mu'min. 1999. "The USDA: Trying to Perpetrate the Perfect Crime." *The Farmer* 2: 8. URL: http://www.muhammadfarms.com/Perfect%20Crime.htm.

Mutz, Kathryn M., Gary C. Bryner, and Douglas S. Kenney (eds.). 2002. *Justice and Natural Resources: Concepts, Strategies, and Applications.* Washington, D.C.: Island Press.

Nabhan, Gary Paul. 1997. *Cultures of Habitat.* Washington, D.C.: Counterpoint.

Nader, Laura. 1996. *Naked Science: Anthropological Enquiry into Boundaries, Power, and Knowledge.* London: Routledge Press.

Nostrand, Richard. 1992. *The Hispano Homeland.* Norman: University of Oklahoma Press.

Oliver, Melvin L., Michael Conroy, E. Walter Coward Jr. 2000. "Building Natural Assets and Wealth to Reduce Poverty: Concepts and Potential Applications in the United States." Paper prepared for the Natural Assets: Democratizing Environmental Ownership conference, Santa Fe, New Mexico.

Parry, B. T., Hank J. Vaux, and Nick Dennis. 1989. "Changing Conceptions of Sustained-Yield Policy on the National Forests," in *Community Stability in Forest-Based Economies,* Dennis LeMaster and John Beuter (eds.), proceedings of a conference in Portland, Oregon, November 16–18, 1987. Portland: Timber Press.

Parson, Edward A., and William C. Clark. 1995. "Sustainable Development as Social Learning: Theoretical Perspectives and Practical Challenges for the Design of a Research Program," in *Barriers and Bridges to the Renewal of Ecosystems and Institutions,* Lance H. Gunderson, C. S. Holling, and Stephen S. Light (eds.), pp. 428–460. New York: Columbia University Press.

Pinchot, Gifford. 1947. *Breaking New Ground.* New York: Harcourt, Brace and Company.

Pipkin, Jim. 1998. *The Northwest Forest Plan Revisited.* Washington, D.C.: U.S. Department of the Interior, Office of Policy Analysis.

Plumwood, Valerie. 1998. "Inequality, Ecojustice, and Ecological Rationality," in *Debating the Earth: The Environmental Politics Reader,* John Dryzek and David Schlosberg (eds.), pp. 559–583. New York: Oxford University Press.

Prugh, Thomas. 1995. *Natural Capital and Human Economic Survival.* Solomons, Maryland: ISEE Press.

Qi, Yadong, Jammie Favorite, and Alfredo Lorenzo. 1998. *Forestry: A Community Tradition.* A joint publication of the USDA Forest Service, the National Association of State Foresters, Southern University, and A&M College. Washington, D.C.: National Association of State Foresters.

Rasker, Ray, and John Roush. 1996. "The Economic Role of Environmental Quality in Western Public Lands," in *A Wolf in the Garden: The Land Rights Movement and the New Environmental Debate,* Philip Brick and Cawley McGregor (eds.), pp. 185–206. Lanham, Maryland: Rowman & Littlefield.

Richard, Tim, and Sam Burns. 1998. "Beyond 'Scoping': Citizens and San Juan National Forest Managers, Learning Together." *Journal of Forestry* 94(4): 39–43.

Ringgold, Paul. 1998. *Land Stewardship Contracting in the National Forests: A Community Guide to Existing Authorities.* Washington, D.C.: Pinchot Institute for Conservation.

Robbins, William. 1988. *Hard Times in Paradise: Coos Bay, Oregon, 1850–1986.* Seattle: University of Washington Press.

Robbins, William. 1989. "Lumber Production and Community Stability: A View from the Pacific Northwest," in *Community Stability in Forest-Based Economies,* Dennis LeMaster and John Beuter (eds.), proceedings of a conference in Portland, Oregon, November 16–18, 1987. Portland: Timber Press.

Romm, Jeff. 2000. "The Social Diversification of Forests." Paper presented at the Cultures and Biodiversity Congress. Kunming, China. July.

Romm, Jeff. 2002. "The Coincidental Order of Environmental Justice," in *Jus-*

tice and Natural Resources: Concepts, Strategies and Applications, Kathryn Mutz, Gary Bryner, and Douglas Kenney (eds.), pp. 117–137. Washington, D.C.: Island Press.

Ruth, Lawrence, and Richard Standiford. 1994. *Conserving the California Spotted Owl: Impacts of Interim Policies and Implications for the Long-Term.* Davis: University of California Wildland Resources Center.

Sauer, Carl. 1938/1963. "Themes of Plant and Animal Destruction in Economic History." *Journal of Farm Economics* 20: 767–775. Preprinted in Carl Sauer's *Land and Life* (1963). Berkeley: University of California Press.

Schumacher, E. F. 1973. *Small Is Beautiful: Economics as if People Matter.* New York: Harper & Row.

Selznick, Philip. 1984 (1949). *TVA and the Grass Roots: A Study of Politics and Organization.* Berkeley: University of California Press.

Sen, Amartya. 1999. *Development as Freedom.* New York: Knopf.

Shepherd, Jennifer. 2001. "Detroit Block Clubs Apply Social Forestry." *Communities and Forests* 5(1): 6.

Sherraden, Michael. 1991. *Assets and the Poor: A New American Welfare Policy.* New York: M. E. Sharpe.

Shindler, Bruce, and Lori A. Cramer. 1999. "Shifting Public Values for Forest Management: Making Sense of Wicked Problems." *Western Journal of Applied Forestry* 14(1): 28–34.

Shutkin, William. 2000. *The Land That Could Be: Environmentalism and Democracy in the Twenty-First Century.* Cambridge, Massachusetts: MIT Press.

Silva, Nattie, Sunny Hemphill, Suraj Ahuja, Dorothy Stennis, Jenny Stephenson, Amahra Hicks, Joan McDevitt, and Curt Peterson. 1988. *Toward a Multicultural Organization (TMO) Revisited: An Evaluation Report and Building Blocks to Meet 21st Century Challenges.* Washington, D.C.: U.S. Forest Service.

Stegner, Wallace. 1953. *Beyond the Hundredth Meridian: John Wesley Powell and the Second Opening of the West.* New York: Penguin.

Stewart, William. 1996. "Economic Assessment of the Ecosystem," in Sierra Nevada Ecosystem Project: Final Report to Congress, Volume III, *Assessments, Commissioned Reports, and Background Information,* pp. 973–1063. Davis: University of California, Centers for Water and Wildland Resources.

Sutter, Paul. 1999. "A Retreat from Profit: Colonization, the Appalachian Trail, and the Social Roots of Benton MacKaye's Wilderness Advocacy." *Environmental History* 4(4): 553–577.

Tandon, Rajni. 1998. "Social Transformation and Participatory Research." *Convergence* 21: 5–18.

Taylor, Frank. 2001. Testimony to Forest Service, Washington, D.C.

Tilly, Charles. 1994. "Social Movements as Historically Specific Clusters of Political Performance." *Berkeley Journal of Sociology* 38: 1–30.

Toulmin, Stephen. 2001. *Return to Reason.* Cambridge, Massachusetts: Harvard University Press.

Umholtz, Justin, and Beverly Brown. 2002. Are Forest "Brush" Harvesters Employees? A Report on the Current Status of the Debate and Its Relationship to

National Issues in Non-Timber Forest Products. Jefferson Center Bulletin 4. Wolf Creek, Oregon: Jefferson Center for Education and Research.

USDA and USDI. 1994. "Record of Decision for Amendments to Forest Service and Bureau of Land Management Planning Documents Within the Range of the Northern Spotted Owl: Standards and Guidelines for Management of Habitat for Late-Successional and Old-Growth Forest Related Species Within the Range of the Northern Spotted Owl." Washington, D.C.: U.S. Government Printing Office 1994–111/00001 Region no. 10.

Viederman, Stephen. 1996. "Sustainability's Five Capitals and Three Pillars," in *Building Sustainable Societies: A Blueprint for a Post-Industrial World,* Dennis Pirages (ed.), pp. 45–54. Armonk, New York: M. E. Sharpe.

Walters, Carl J. 1986. *Adaptive Management of Renewable Resources.* New York: Macmillan.

Walters, Carl J., and Crawford S. Holling. 1990. "Large Scale Management Experiments and Learning by Doing." *Ecology* 71: 2060–2068.

Walzer, Michael. 1982. "Politics in the Welfare State: Concerning the Role of American Radicals," in *Beyond the Welfare State,* Irving Howe (ed.), pp. 129–154. New York: Schocken.

Weber, Edward. 2000. "A New Vanguard for the Environment: Grass-Roots Ecosystem Management as a New Environmental Movement." *Society and Natural Resources* 13(3): 237–259.

Wellman, J. Douglas, and Terence J. Tipple. 1990. "Public Forestry and Direct Democracy." *The Environmental Professional* 12: 76–85.

Williams, Bruce, and Albert Matheny. 1995. *Democracy, Dialogue, and Environmental Disputes.* New Haven, Connecticut: Yale University Press.

Williams, Raymond. 1980. "The Idea of Nature," in *Problems in Materialism and Culture: Selected Essays,* pp. 67–85. London: Verso Press.

Wolfe, Alan. 1995. "Social and Natural Ecologies: Similarities and Differences," in *Seedbeds of Virtue: Sources of Competence, Character, and Citizenship in American Society,* Mary Ann Glendon and David Blankenhorn (eds.), pp. 163–183. New York: Madison Books.

Wollenberg, Eva, David Edmunds, and Jon Anderson (guest eds.). 2001. "Accommodating Multiple Interests in Local Forest Management," in *International Journal of Agricultural Resources, Governance and Ecology* 1: 3/4. Oxford, England: Information Press.

Yaffee, Steven L. 1994. *The Wisdom of the Spotted Owl: Policy Lessons for a New Century.* Washington, D.C.: Island Press.

Young, Iris Marion. 1990. *Justice and the Politics of Differences.* Princeton, New Jersey: Princeton University Press.

Zack, Michael H. 1999. "Managing Organizational Ignorance." *Knowledge Directions* 1: 36–49.

About the Authors

Mark Baker is a research associate with Forest Community Research based in Arcata, California. His primary interests concern community-based resource management. His work has focused on social forestry and community irrigation systems in India and watershed institutions and community forestry in the United States. He received his Ph.D. in wildland resource science from the University of California at Berkeley.

Jonathan Kusel is founder and director of Forest Community Research, where he also directs the Pacific West Community Forestry Center. He holds a Ph.D. in natural resource sociology and policy from the University of California, Berkeley. His research focuses on community health and development, socioeconomic monitoring and assessment, and community forestry.

Jeffrey G. Borchers is a researcher, writer, and consultant living in Jacksonville, Oregon. He is a research forester with Pacific Southwest Research Station, USDA Forest Service. Dr. Borcher's works bring a systems perspective to both the social and scientific dimensions of sustainable ecosystem management, particularly community forestry. Current activities include the creation of public–science–agency partnerships for adaptive ecosystem management and the design and implementation of decision–analytic processes to support collaborative environmental planning, decision making, and monitoring.

Leah Wills is research associate with Forest Community Research in Taylorsville, California. She received an M.A. in rural development from California State University, Chico. She works on water and water-related issues, including rural water rights protection, source water quality improvement, and on securing investment into the "areas of (water) origin" for land and water stewardship.

Index